The Command Decisions Series

• VOLUME 4 •
Instrument Operations

The Command Decisions Series • Volume 4

Instrument Operations
Air Safety Considerations During IFR Flight

Richard L. Taylor

Belvoir Publications, Inc.
Greenwich, Connecticut

Also by Richard L. Taylor

IFR for VFR Pilots: An Exercise in Survival
Understanding Flying
Instrument Flying
Fair-Weather Flying
Recreational Flying
Positive Flying (with William Guinther)
The First Flight
Pilot's Audio Update (Editor)

ISBN: 1-879620-05-7

Printed and bound in the United States of America by Arcata Graphics (Fairfield, Pennsylvania).

Contents

Preface

Instrument flying is taken for granted by a considerable number of pilots in today's aviation community. Airline and business-aviation operations are shut down by only the most severe weather conditions; approaches and landings in *very* low visibilities have become routine with the advent of high-technology equipment, and weather delays stem from lack of capacity in the airspace system rather than the pilots' inabilities to move their machines through inclement conditions. This segment of aviation is nurtured and enhanced by pilots with high experience levels and good training, flying the best equipment, and almost universally operating with two-pilot crews—a cockpit assistant and a "second opinion" is frequently an invaluable asset when the chips are down.

But as usual in this topsy-turvy world of flying, those who need the most help have the least; inexperienced pilots flying IFR by themselves in light airplanes with minimum equipment are nevertheless permitted to operate in whatever weather conditions they choose to challenge. Most of the time—almost *all* of the time, as a matter of fact—instrument flights are completed safely, but now and then there's a miscue, a moment of inattention, a faulted procedure, an emergency situation which results in an unscheduled contact with the ground. It's those relatively few—but frequently fatal—accidents that occur during IFR operations that we will address in this volume of *Command Decisions*.

Our objective is to help you understand why these accidents took place and thereby help you avoid making the same mistake. Perhaps even more important, you may someday recognize a similar situation setting up, and make a change in time to forestall an accident. Remember that we *must* learn from the mistakes of others—we will not live long enough to make them all ourselves.

Richard Taylor
Dublin, Ohio
July 1, 1991

Instrument
Training

IFR proficiency is largely a matter of perception. A pilot might jump marvelously through all the training hoops on the way to an instrument rating, and make a very favorable impression on the examiner; in this case, the CFI and the evaluator perceive that here is one sharp IFR airplane driver, and the pilot would no doubt agree.

On the other hand, an instrument trainee might be something less than super-sharp on the gauges, and the examiner advises him that although the pilot meets the minimum standards for the rating, he must always be content to operate in conditions that provide margins for his skill level.

In either case, personality comes into play. The first pilot might be so pumped up by his performance during training and on the checkride that he considers himself capable of handling anything the IFR system or Mother Nature can throw at him; the second pilot might recognize his shortcomings and not only respect his personal limitations, but wisely continue with regular training to keep whatever edge he can attain. A pilot's perception of his own proficiency is the most important.

Regulatory Restraint...Almost Nil

The minimum standards prescribed in Part 61 for an instrument rating are relatively liberal, and are open to considerable qualitative interpretation by examiners. To be sure, there are finite boxes into which an applicant's performance must fit, but the monstrous unknown is *judgment*; even though a pilot demonstrates respect for the regulations and procedures on the checkride, will he continue with that healthy attitude when he gets out in the system by himself? It's the toughest call an examiner must make.

And Part 91 doesn't help a lot. The operating rules for a non-commercial flight permit a departure in zero-zero conditions for any pilot with a current instrument rating, and on the other end of a flight, permit an approach regardless of visibility. The increased workload and proficiency requirements of an approach to minimums (or below) have led some aviators to conclude that we should adopt a set of graded IFR privileges; higher MDAs and increased visibility requirements until certain levels of experience have been acquired, for example. But all attempts at changing the rules in this direction have come to naught, and it remains up to each Part 91 pilot to determine his own minimums.

Despite the keen edge of competence a pilot has after gaining his instrument rating, few pilots ever immediately go out and fly approaches to minimums, and many arrange their missions so that IFR flying at first only involves the en route portion of a flight.

There must come a time, however, when the pilot will have both the need and the confidence to shoot his first "actual" approach, and eventually to adopt the published minimums as his guidelines. When does the day come? Is it in the first 25 hours of IFR flying, the first 50, the first 100? And when the day arrives, will the pilot be in circumstances where he feels compelled to attempt the approach, or will he have researched the situation so that he has an alternate well above minimums in case he doesn't feel up to the task?

These questions appear germane in the crash of a Cherokee Six at Fort Wayne, Indiana. All four occupants died when the plane crashed just abeam the runway threshold on a night ILS approach to Runway 4. The pilot had gained his instrument rating 10 months before the accident; at that time his log showed a total of 412 hours, 77 of which were instrument instruction. His logbook found at the time of the crash showed a total of 98 instrument hours, 25 of them actual, but no instrument time within the 90 days prior to the crash.

Before departure from a Florida airport, an alternator fluctuation squawk was fixed by tightening a loose connection. The Cherokee Six flew to Athens, Georgia, where the tanks were topped off, the pilot got a weather briefing and filed IFR for Fort Wayne.

Although they are restrictive for for-hire operators, the FARs allow Part 91 operators a surprising amount of latitude.

The cruise portion of the flight at 8,000 feet was mostly routine, although at around 9 p.m. a Cincinnati controller noticed the plane making a turn to the south, and questioned the pilot. "Yeah, this instrument apparently messed up. I had it on autopilot. I'm getting back on course," the pilot said.

When he contacted Fort Wayne Approach Control at 9:50 p.m., the pilot was given the current altimeter and a weather observation: "Indefinite ceiling zero, sky obscured, visibility one-eighth mile, light drizzle and fog." The pilot responded, "Okay, we'll shoot the ILS Runway 4 and see if we can get down."

The pilot had filed Muncie, Indiana, as his alternate (he would nearly overfly it on the way to Fort Wayne). Less than a minute after requesting the Fort Wayne ILS approach, he asked the controller, "What's the weather like in Muncie if we have to go into our alternate?" The controller checked and responded, "Muncie weather is identical to ours."

In due time, the pilot started his descent, and was

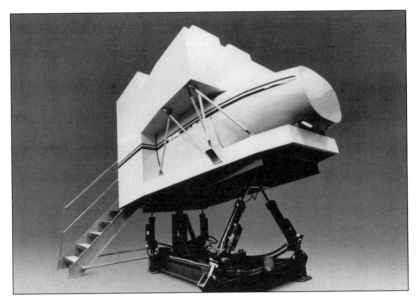

Simulator training, such as that offered by FlightSafety International, is an expensive but effective way to keep IFR skills sharp.

cleared to 5,000 feet about 30 miles south of the airport. Along the way, the controller announced that the Runway 4 RVR was 1,400 feet, and the pilot acknowledged.

Seven minutes later, the pilot requested a lower altitude because "we're starting to pick up quite a bit of ice." He was immediately given clearance to 2,800 feet, which is roughly the initial approach altitude.

The controller asked the type and degree of icing, and the pilot called it "rime" and "moderate." Four minutes later, the controller asked about the icing conditions and the pilot said, "We've still got ice but it doesn't seem to be building any more."

The flight was vectored for the approach and started down the ILS course. The pilot reported the outer marker inbound (6.5 miles from the runway) and was handed off to the local controller, who reported, "Prevailing visibility one-eighth of a mile, RVR Runway 4, 1,600 feet, cleared to land."

As far as the controllers could tell, the Cherokee Six stayed on the localizer and glide path until it was a mile and a half short of the runway, when the controllers noticed significant deviations in both course and altitude. Radar showed the Cherokee descended to field elevation, and when there were no responses to repeated radio calls, the controllers called out search and rescue teams to look for the downed airplane.

Physical evidence suggested that the plane flew into a 55-foot tree on a heading of about 40 degrees, struck the ground 560 feet later, bounced and went 80 feet before coming down again and beginning to slide and break up. This and other evidence is consistent with an initial impact in roughly level flight with the engine running. The crash came just abeam and one-fifth of a mile to the right of the runway threshold.

Unfortunately, the wreckage did not provide enough clues to establish whether the plane might have been in a shallow glide, level flight or starting a go-around, nor could the avionics and instruments give a coherent picture. The ILS was checked and found to be in perfect working order. Evidence of airframe ice was not documented by investigators (the temperature on the ground was 36 degrees).

It probably will never be determined whether the pilot had some navigation problem or failure, thought he saw something like the runway environment, or even started a missed approach but didn't get a chance to complete it. But it is certain that he commenced the approach in the face of weather below minimums, and struck a tree in a place where he should have been 150 feet higher.

While this relative novice made a risky stab at getting into Fort Wayne, experienced professionals were executing missed approaches and diverting elsewhere. In the hour before the Cherokee Six's attempt, a Mooney M20E in the hands of a 10,000-hour ATP missed one approach and made another; an Air Wisconsin jet missed two attempts (ceiling was reported as 100 feet and RVR 5000 at the time) and diverted to Toledo, Ohio; another Air Wisconsin flight, hearing the weather observation, broke off and diverted to Lafayette, Indiana, about 10 minutes before the Cherokee Six contacted approach control.

It is interesting to note that both airline crews found weather above minimums at their alternates, and these airports were both within range for the lightplane, which had well over an hour's worth of fuel left. Had the pilot made the request, the controllers no doubt could have suggested several airports where conditions were a lot better than at Fort Wayne.

Perhaps the epitaph was contained in the statement given by the Mooney pilot, whose first attempt was missed after he saw part of the approach light system but not enough to satisfy his standards. Gaining full view of the approach lights on the second try, he was able to land, about an hour before the Cherokee Six arrived. Said the Mooney pilot, "In my judgment, weather conditions at the time of my landing were right at minimums for a 'raw data' approach flown by an experienced and current instrument pilot."

Jim and Mary and Herb and Tanya

Sorry, this has nothing to do with party games. In fact, this foursome has never met. The only thing the four have in common is a pilot license with instrument rating, and particularly hairy experiences with approaches at near-minimums.

Jim owns an older single-engine, four-place retractable. It's been completely refurbished, including just about every avionics marvel known to man. He uses the airplane primarily for business, and seldom cancels for weather. But he gets very little approach practice—most of his "instrument currency" time is en route, and he logs every instrument approach procedure as "actual," even if he breaks out of the gook at 3,000 feet AGL.

Today's appointment is an important one, and he really has to get there, even though his arrival area is right in the center of widespread low ceilings and fog, expected to last for several more days. Three different airlines at his home field offer flights to his destination, but Jim doesn't consult them—after all, he is "current" on instruments.

As he gets close to his destination, reported weather matches the terminal forecast closely: "indefinite ceiling, 400 feet obscured, one-half mile in fog and light rain." Jim accepts a clearance for the ILS Runway 5. He is still looking over the approach plate as the ADF needle swings, the marker beacon sounds and the glideslope needle moves down through center. He doesn't punch his stopwatch at the FAF because he is still trying to memorize Decision Height as the FAF falls behind.

Jim really should abandon his approach now—he is all fouled up. But he continues, chasing a full-fly-down indication on the glideslope and a three-dot localizer needle deflection; in fact, he is

The glideslope may be a dot high or low but full-scale deflection of the localizer is cause for an immediate missed approach.

still chasing both needles at DH. With nothing but a housing development in sight, he wisely executes a missed approach. Things are okay, but he will need a fresh shirt and skivvies before his business meeting.

The tower reports, "We saw you going by the south end of the field on your missed approach." But Jim never saw the field, and for good reason; he had put in a big 20-degree heading correction to deal with a sudden full-scale localizer needle deflection while he was trying to correct a big "fly-up" command on the glideslope. At DH, he looked straight ahead—not to the left where he would have seen the runway.

Jim tried another approach, and made it. This time he was worried that his glideslope receiver was bad, so he elected to do the localizer-only procedure. This time he punched his stopwatch at the FAF, and had thoroughly memorized the MDA. In fact, he flew the whole approach on the No. 2 nav, which had only a localizer indication—no glide slope. And it was a nice approach, on centerline all the way. The gear went down at the FAF, and he saw the runway from MDA before passing the middle marker, in time to extend full flaps and make a normal landing right on the spot.

The moral? The "minimums" on the approach plate are for "proficient" pilots; "instrument currency" doesn't equate to "instrument proficiency." Shooting the ILS to minimums requires the ultimate in "proficiency." Instead of just adding 200 feet to your personal minimums when you know you are out of practice, why not opt for the localizer approach the first time, level off at the safer MDA, and if you don't see the runway at the middle marker or by the end of your timing, just go to the alternate?

Jim's second approach was much better because he was ready for it, and he didn't have to chase the glideslope. His preparation enabled him to set power for the desired rate of descent, and concentrate on airspeed/power/localizer and the MDA.

Jim's experience doesn't sound too hairy, does it? Changing your shirt after a long flight in a non-air-conditioned airplane isn't too unusual if you want to impress your clients. But why did he have to change his skivvies? Because when Jim reached for the gear handle during his missed approach from the ILS, *the gear handle was already up!*

Enter a Strange Airplane

Mary is a lot more proficient than Jim—at least in ILS approaches. Once each month she goes up with a flying buddy and practices ILS and VOR approaches under the hood. She is even pretty good on ADF, and can intercept and track a mean inbound or outbound track. Usually she gets practice in her boss's big twin which is fully equipped, including a slaved RMI.

But today she is on personal business, and has to rent a light twin that she has never flown before. She is flying up into the New England hills to see friends and like Jim, has to face low ceilings and rain. The only approach available is an NDB procedure, and the beacon is several miles northwest of the single-runway airport.

Mary is well prepared for this approach. She knows the MDA, the final approach bearing, and the missed approach procedure. She has reviewed the procedure long before being cleared for the approach. She has checked the DG against the wet compass while outbound for the procedure turn, and has reset the DG after the procedure turn to cancel out a small disagreement between the two instruments.

Everything goes well. She breaks out of the overcast above the

The outcome of an NDB approach is only as good as the accuracy of the compass. Correction cards should be checked and kept up to date.

MDA, but still has about two miles to go to the airport and can't see much through the rain-covered windshield. Suddenly there's a razor-back ridge covered with pine trees, *dead ahead!*

Mary's missed approach is a little more accentuated than Jim's—she has to pull up over rising ground—but it's a safe exit from near-death. Mary flies to her alternate, and calls her friends to come pick her up—in an automobile.

The problem? It was not entirely Mary's fault. Her rented airplane had a badly erroneous wet compass, with its worst error on southeast headings. The deviation card showed near-zero error on all headings when in fact, the error on the final approach heading was nearly 20 degrees—more than enough to offset the final approach segment directly into the hills.

The moral? An NDB approach is only as good as the magnetic heading information available to the pilot. If you don't know the airplane, better not trust the magnetic compass until you have checked it yourself.

ILS Proficiency Is Great, But...

So how about Herb? He is a lot more prepared than Jim, and about on a par with Mary. Herb shoots a neat ILS, and knows not to put in big corrections as he nears minimums. He will accept a one-dot localizer or glideslope error. But he has not flown a simple VOR-DME approach for a long time. Today, he has to do one—possibly two or three of them. He plans to meet some golf buddies at an outlying airport 12 miles away from the major field with the VOR and an ILS.

On the first two approaches he reaches the 12-mile DME missed-approach point way below the overcast, but can't see the airport because of the fog and haze. He's never more than one, maybe 1.5 dots off on the CDI. Herb has plenty of gas, so he tries one more. This time he is right on the money with the CDI, and at 12 DME he has a runway just under his left wing.

He bends a hard left, chops it to idle, remembers the gear but forgets the flaps, and lands *hard* on Runway 4. Unfortunately, he was cleared to land on Runway 9.

Herb's problem? It so happens that his No. 1 VOR indicator was borderline on its four-degree maximum allowable error, and on his first two approaches his own navigational error was in the same direction due to a left crosswind. At 12 miles from the VOR, this gave him about a six-degree total track error, put him a mile south of the desired course.

Luckily for Herb, there was no obstruction southwest of the airport. Also, luckily for Herb, his golfing buddies were not aviators. In fact, they were really impressed with his flying: "Gosh, Herb, we heard you go by two times, then about a half-hour later, there you were, taxiing in. Really cruddy weather. You must be a good pilot."

It took the local avionics shop only half an hour to calibrate Herb's VORs and eliminate the excessive error. He should have been performing regular VOR checks (the regulations require them every 30 days), and should have had the maintenance done a year earlier. And of course, on a long VOR-DME approach, he should have been proficient enough to fly *zero* CDI error.

Proficiency Is as Proficiency Does

What about Tanya? She was ferrying a medium-sized recip twin back home for maintenance on the right alternator. Fifty miles from her destination, the right engine started losing oil pressure, along with high oil temperature. Tanya would not have accepted this flight into IFR weather if she had not had considerable experience in this type of airplane. She knew the bird would fly acceptably on one engine at its low ferry weight. And she knew what power settings and configurations to use for an engine-out instrument approach.

Tanya declined radar vectors to a nearby VFR-only airport, because she didn't want to operate VFR under a low ceiling and in poor visibility. She also declined her original destination which had only a VOR approach, knowing she would have to circle-to-land into moderate winds at that airport. Instead, Tanya changed her destination to a field a little further ahead that had full ILS and a full-time crash crew.

When the right oil pressure dropped to a marginal value, Tanya shut the engine down and feathered the prop. She already knew what power to set on the left engine for level cruise and glidepath descent. She already knew that she could hold the ILS glidepath with the gear down, zero flaps, and less-than-full power on the operating engine.

She made a point of staying at or slightly above V_{yse}, and on or slightly above glidepath, and because she knew the performance of this airplane, she was ready to do a single-engine go-around if necessary, as long as it began at least 500 feet AGL. Tanya shot a near-perfect ILS the first time, breaking out just above minimums and making a good landing. Thinking ahead, Tanya had even told the tower she would need a tow to the ramp.

How could Tanya make a good approach and landing—even with a single-engine emergency—when the rest of our quartet scared the heck out of themselves or others? Because Tanya had reserves of currency and proficiency. Tanya had thought about each possible problem and circumstance ahead of time. In a word, Tanya was *prepared*.

Recovering from Unusual Attitudes

Test pilot and flight instructor Bill Kelly has gotten himself into and recovered from every unusual attitude imaginable. Here's his view of how this important skill should be developed and honed..

Imagine that you and your CFII are flying along straight and level, then, after a quick warning from your instructor, you find yourself in a steep bank, with the nose 30 degrees below the horizon.

Recover! Try not to exceed Vno (maximum operating speed). By all means, stay below Vne (never exceed speed). Don't exceed the airplane's G limits. Don't let the engine and prop overspeed.

A pilot is often presented with that scenario—or something very similar—during a Flight Review. Typically, there's a pre-briefing on recovering from unusual attitudes. The air work is often done with the attitude and direction indicators covered—a worst-case situation.

Speed Kills

One of the most common errors is failing to adjust the throttle, even when the nose is low and airspeed is increasing. That power lever has got to come back to idle, right now. To heck with concerns about shock-cooling the engine; if airspeed is building rapidly, power has to be cut quickly.

Just the opposite holds, of course, if it's a nose-high unusual attitude. Go to full power pronto, and try to keep the airspeed above the 1G stall value. Keep the angle of attack below stall angle by easing forward on the control wheel—almost to zero G.

If a nose-high, decreasing airspeed situation is not handled promptly and correctly, it will likely turn into a graveyard spiral or a spin. Recovery will certainly be from a nose-low, high-speed condition that could shed the wings.

So, how do you make that high-speed recovery? Reduce power and level the wings. Don't try to get the nose back up to the horizon until bank has been reduced to zero. Use coordinated aileron and rudder pressure to roll the wings level. You might want to hold a little forward pressure on the control wheel so that G load doesn't build during the roll-out.

A graveyard spiral could develop if you try to get the nose up while still in a steep, descending turn. Up-elevator may only increase the G loading and speed as the nose drops even further. Bank angle may get even steeper. Airspeed, G load and engine rpm (with a fixed-pitch prop) will go past all limits. The wings and parts of the empennage might be found by investigators far from the big hole dug by the fuselage and engine.

Be careful too, that you don't roll rapidly out of the steep descending bank while you're pulling high G at a high airspeed. That's called a "rolling pullout," and *really* loads the wing. But very often, that's just what happens when even experienced pilots recover from one the relatively mild nose-low unusual attitudes set up for them in training.

Discussions of unusual attitudes usually evoke casual dismissals, such as "no decent pilot would let his airplane get into such a situation," or "a good instrument pilot flies attitude and never gets into steep banks." Sure—but lots *do* crash each year from out-of-control turning descents. We are talking about the real world here.

Spiral Divergence

That real world includes a phenomenon that can be exhibited by any airplane—when it gets beyond a particular bank angle and won't tend to roll back to wings-level all by itself, that's spiral divergence.

An airplane in a shallow left turn tends to carry a little left sideslip angle. The nose isn't quite following the turn, and the relative wind is slightly from the left side—vice versa for right turns, of course. Light airplanes have a small amount of effective dihedral, created primarily from the angle at which the wings attach to the fuselage and/or by a swept leading edge.

That effective dihedral provides a rolling moment away from the sideslip; it tends to roll the plane back to level from a shallow bank. Dihedral effects assist you in holding the wings level, and permit you to make shallow turns with just the rudder pedals. But once you get past 10 to 20 degrees of bank, the airplane probably won't recover to wings-level by itself. Past about 30 degrees, most airplanes will tend to *increase* the bank angle with no input from the pilot, and we are into spiral divergence.

The turn radius is decreased sufficiently that the outboard wing is moving with a measurably higher airspeed than the wing on the inside of the turn, and because of the unbalanced lift, the airplane wants to roll even steeper. If you haven't added sufficient up-elevator to maintain a vertical component of lift equal to weight, you start to descend.

That increasing descent in the left turn means that initially the airplane is also slipping left. Now, directional stability is stronger than dihedral and the airplane wants to point its nose into the relative wind. So the nose yaws left. But in this steep left bank, left is also *down;* the nose drops and speed increases. And even without any additional up-elevator, G load increases due to the effect of the elevator or stabilator trim setting.

Now, it's probably too late to recover by using only up-elevator. Any nose-up command is only going to tighten the turn, increase the G loading, allow the nose to drop further and cause the airspeed to build even more. You must roll the wings level and get the power to idle before the wings are torn off or, if it's nighttime or you're in clouds, before the attitude indicator tumbles.

Deadly Slowdown

Now, let's consider some scenarios that could set up a graveyard spiral. Imagine an inexperienced pilot or one who's current on instruments but not really proficient and needs work on his scan. Take your pick of the conditions; a dark night, IMC, daylight with haze obscuring the horizon, or even worse, haze over calm water.

The Bugbuster normally cruises at 120 knots and stalls at about 60. The pilot is concerned about turbulence or is trying to dodge clouds in the hazy murk. Maybe he's lost and decides to conserve fuel while he searches for a recognizable landmark—in any case, he reduces power and retrims for 90 knots.

That slowdown could be deadly.

Our pilot gets real busy reading charts, talking on the radio or tuning the nav radios. He trimmed the elevator for 90 knots but didn't trim the rudder. (If the Bugbuster is a single-engine airplane, it probably doesn't have rudder trim, but is rigged for straight flight at normal cruise speed and power.) While the pilot is busy in the cockpit, the left wing slowly begins to drop. There's no clear horizon, and the change doesn't register on his peripheral vision. A spiral is beginning, and the nose-up trim is going to get him into trouble.

He thought he had trimmed for an airspeed—90 knots—but in reality, trim (just like the elevator itself) changes the angle of attack. Sure, his nose-up trim sets up the airplane for the desired airspeed, but only in wings-level, 1G flight. Once the spiral starts, the higher angle of attack is going to cause a tighter turn, more G, more airspeed, and may cause the wing to separate from the fuselage. There's a good chance that the low-speed, level trim is good for 6 Gs at 200 knots in a runaway spiral—and without any further elevator input from the pilot.

Next time you're up for some air work, set up the airplane for slow flight, then take your hands off and let it wander. Put a little pressure on a rudder pedal, as if you inadvertently did so while reaching for something in the back seat. Watch how quickly that spiral develops, how rapidly airspeed, bank angle, and G increase, even with both hands and feet off the controls.

Before speed and G load get excessive, roll back to wings-level and note how fast the nose pitches up way above the horizon because of the trim. A nose-high unusual attitude might mean a stall and spin.

The moral? Don't trim for slow flight unless you really have to, and only when you can devote full time to the instruments or a visual horizon.

The purpose for reducing power if you find your airspeed building rapidly is to slow the rate of buildup. But suppose you fly behind a fixed-pitch prop? You had better be real concerned about not flinging a prop blade—or the engine. Most fixed-pitch propellers nearly reach redline RPM at full-throttle airspeed in level flight. They also reach redline RPM just above redline airspeed with the throttle at idle.

Dump Your Lift

Let's get back to a subject we've skirted—nose-high unusual attitudes. Unless you're practicing aerobatic maneuvers or let a nose-low recovery take itself nose-high or mishandle a big power addition in an airplane that's very sensitive to such a change, there's no tendency for a nose-high divergence.

When you find yourself in an unusual attitude with the nose very high and airspeed decreasing, forget the existing bank angle; don't try to level the wings just yet, but go to full power and move the control wheel forward. Dump your lift. Reduce the angle of attack. Make yourself light in the seat as you reduce G loading to almost, but not quite, zero.

The Bugbuster stalls at 60 knots in level flight at max gross weight. If almost all of the wing lift is dumped, the Bugbuster's ailerons, elevator and rudder probably will still remain effective down to 20 or 30 knots; if the push on the wheel is enough to lower angle of attack way below the value for maximum coefficient of lift, the Bugbuster probably won't stall and spin.

But don't use a lot of aileron in an attempt to level the wings while the nose is still way up and airspeed is still very low. Adverse yaw from aileron deflection might start a big yaw rate that could lead to a spin.

Spin Recovery

Now let's consider another unusual attitude—the final phase of recovering from a spin, when autorotation has stopped, the nose is pointing almost straight down and airspeed is building quickly.

Recovery from this unusual attitude is often mishandled, even by experienced pilots. Most pilots tend to pull too little G initially, and the airspeed gets way too high.

You have to get the wings level and apply sufficient G while the airspeed is still within bounds. Get the pitch-up to level flight started—and hopefully completed—while the airspeed is still within the green arc.

Your *Pilot's Operating Handbook* may include a statement such as "Recovery from an inadvertent one-turn spin may consume as much as 1,000 feet of altitude." Would you be surprised to find out that the test pilot who did those certification spins probably used about 3 Gs during the dive recovery? An airplane certified in the Normal category can easily be recovered from a one-turn spin without exceeding the G-load limits, but it's very easy to overstress the airframe and overspeed the engine on recovery from the ensuing dive.

Some airplanes are certified in both the Normal and Utility categories, with spins permitted when operated according to the requirements for Utility work. Many of these airplanes don't really want to stay in a spin when they're within the required weight and balance limits—they tend to "self-recover." But what if the pilot tries for a four-turn spin by holding the elevator all the way back and full rudder into the spin?

First, it's easy to miss just when the spinning stops unless you are watching some special instruments. They'll show airspeed starting to build rapidly, an increase in roll rate but a slowing of yaw rate. You can't really see this through the windshield, or with reference to the flight instruments. The airplane appears to be pointing almost straight down and to still be rotating. It is—but mostly about the roll axis and less about the vertical (yaw) axis. A tail-damping stability factor makes full-up elevator less powerful than in non-turning flight, so the airplane may fly unstalled with the flippers full up.

The real tell-tale is the G-meter, but not all Utility airplanes have one. The needle in the G-meter starts to move upwards rapidly as the spin turns into an extreme graveyard spiral. You might go right past Va (maneuvering speed) with full-up elevator and full pro-spin rudder. This is an extremely unusual attitude that puts extreme loads on the airframe.

Don't try even a one-turn spin in a Normal category airplane,

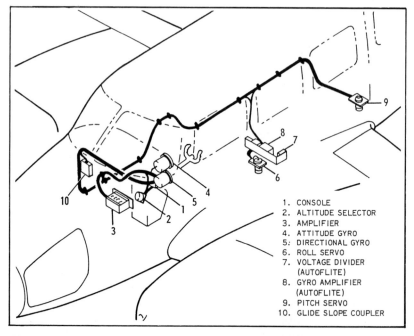

1. CONSOLE
2. ALTITUDE SELECTOR
3. AMPLIFIER
4. ATTITUDE GYRO
5. DIRECTIONAL GYRO
6. ROLL SERVO
7. VOLTAGE DIVIDER
   (AUTOFLITE)
8. GYRO AMPLIFIER
   (AUTOFLITE)
9. PITCH SERVO
10. GLIDE SLOPE COUPLER

Most autopilots are not so complex that the pilot can't master a basic understanding of the important components. Being current on the autopilot's fine points could be a lifesaver if it malfunctions.

and don't mess with spins in Utility or Aerobatic category airplanes without availing yourself of some good instruction.

George's Tricks

Another way you could inadvertently find yourself in an unusual attitude is if, for some reason, something goes awry with your old buddy, "George."

The autopilot has to be watched. It could lose gyro input and roll the airplane past design limits or short-circuit and command full nose-up trim.

Almost any bad command an autopilot comes up with can be over-ridden—but you have to be watching for them and know what to do. Every system has different operating procedures that have to be learned. You have to know—cold—all the methods of disconnecting the autopilot on the airplane you are flying.

You should be able to manually over-ride George without exerting an undue amount of force; but keep in mind that if the autopilot has issued a nose-down command and you belay that order by pulling back on the yoke, you're also running the elevator trim servo towards full nose-down. Then, when you finally disconnect or the elevator servo clutch finally slips, you will be confronted with the full force of the trim tab. Arm-wrestling with George could get you into a wings-level vertical dive past Vne.

Good Exercise

Don't laugh at your CFI's attempts to confuse you with eyes-closed or even eyes-open unusual attitudes; instructors need to see how you react to moderate attitude excursions in order to develop the prompt, rapid and correct responses that you will need if the real situation ever develops.

Don't be insulted if your instrument instructor recommends that you come back for more training on partial-panel procedures before he signs off a Flight Review or an Instrument Competency Check.

Someday, you may be on your own to recover from an inadvertent extreme unusual attitude—and you might have to perform the recovery on partial panel. Only good instruction, followed by lots of practice, can prepare you for that eventuality.

Instruments, Avionics & Aircraft Systems

You've surely heard the story about the world's first completely automated airline flight. The passengers on this robot-controlled operation—no humans in the cockpit—were greeted by a recording which told them to "sit back, relax and enjoy the flight, and rest assured that nothing can go wrong...can go wrong...can go wrong..."

Unfortunately, things do go wrong on occasion, but pilots are trained to recognize and overcome the problems they encounter. Instrument flying is a rather special consideration, because IFR pilots are so totally dependent on the indications of attitude instruments and avionics; the reliability of these appliances is very high, and probably as a result of going for long periods of time with no discrepancies, pilots tend to believe whatever they see on the instrument panel.

Here is some information and case histories that make clear one of the basic precepts of flying—never assume *anything*.

Is Your Altimeter Lying to You?

When you glance at your altimeter, it's likely that you're looking at an instrument manufactured by Tokyo Aircraft Instrument Company and sold by United Instruments. With more than 300,000 units in circulation, United has virtually cornered the low-end altimeter market.

Aside from two ADs, the altimeters have proven to be quite reliable. Many thousands of them are in use daily, and millions of flight hours have been logged with relatively little difficulty. However, Bill Magagnos, the proprietor of a Florida instrument shop believes there's a combination of elements in the design and manufacture of these altimeters that could cause it to hang up during a descent and lead to a crash. He's devised a fix for the alleged problems and has gone so far as to ask the FAA to issue another AD on the altimeter; the FAA and United remain unconvinced that the problem is real or that the fix works.

The evidence does not support the contention that there's an overwhelming problem with United altimeters, but the information is compelling enough to have convinced two juries that altimeter hang-ups contributed to accidents. Both aircraft crashed in instrument meteorological conditions, and in each case there is some evidence that the altimeter was reading hundreds of feet higher than the actual altitude at impact.

The first accident occurred during a localizer approach to Crestview, Florida. The pilot of the Beech Sundowner reported crossing the final approach fix inbound and began descending to the published minimum descent altitude. As the aircraft approached MDA, it suddenly rolled left and pitched down. The pilot, who said he felt a downdraft, recovered and began a turn to initiate the missed approach. A moment later, the airplane hit trees at 200 MSL.

The Sundowner pilot said he saw the altimeter reading 750 feet just before he felt the downdraft, a few seconds before hitting the trees.

The altimeter was removed from the airplane and sent out for testing that showed the instrument was within tolerances. There was a minus 200-foot error in the Kollsman window-setting mechanism, but investigators concluded that it had no bearing on the accident, even though it would have placed the pilot at only 350 AGL at the moment of his last altitude observation. The report also noted that the instrument needed to be overhauled and that it showed some friction.

The NTSB concluded that the pilot was at fault in the accident because he descended below the MDA; the altimeter was not mentioned in the Board's findings.

In the other accident, three people aboard a Bell 206L3 helicop-

As instruments go, the altimeter is as bulletproof as any. Nonetheless, its delicate mechanism has been known to stick, causing higher-than-actual readings.

ter were killed in Pendleton, Oregon. The pilot was instrument-rated in airplanes but not in rotorcraft, even though he had more than 4,200 hours in helicopters.

The weather was marginal, and when the pilot was eight miles west of the Pendleton airport, he contacted ATC and requested a special VFR clearance into the control zone.

Five minutes after getting the clearance, the pilot declared an emergency, saying that he had entered IMC inadvertently and was not IFR-qualified. The helicopter hit the ground at an elevation of 1,435 feet as the pilot was making a transmission: "I'm at 2,200 and I'm 5.1 miles from..."

There was a leak in the pitot/static system that the pilot knew about before the flight. Due to the level of destruction of the helicopter, it could not be determined what effect, if any, the leak might have had on the accident.

Altimeter experts believe the leak could not have accounted for the more than 600-foot altitude discrepancy (nearly 0.6 inches of pressure). The experts say error from such a leak would more likely be on the order of 100 feet or less. But in its determination of probable cause, the NTSB cited the static system leak as a contributing factor. Spatial disorientation of the pilot was deemed the probable cause.

As part of a subsequent investigation, the altimeter was removed from the wreckage and sent to the Florida shop, where it was found that it was sticking in descent. Lawsuits involving both of the accidents went to court, the instrument-shop owner served as an expert witness for the plaintiffs, and the juries found that altimeter hang-ups were significant factors in the accidents.

Inner Workings of the Altitude Meter

The pressure altimeter is a fairly simple mechanical device built around a sealed disc-shaped capsule called an "aneroid" that expands and contracts with changes in air pressure. A brass link about an inch long mounted perpendicular to the face of the disc transmits the motion to a lever, which in turn is attached to the gears that turn the pointers on the altimeter.

In the United altimeter, the brass link has a hole at one end and a slot at the other, with pins through both. During ascent, the aneroid pushes the slot-end pin against the end of the slot. In descent, however, the aneroid is pulling away from the pin. The pin is still held against the end of the slot, but by spring pressure; thus the pin is always positioned firmly at one end of the slot.

The slotted link may contribute to the severity of altimeter hang-ups. If there's enough friction in the mechanism to overcome the spring pressure holding the pin against the end of the slot, the pin can ride up in the slot while the aneroid contracts. Chances of the pin ever reaching the top of the slot would be remote; the airplane would have to descend 11,000 feet before that could happen.

Central to this argument is the amount of force available to drive the gear train. The up-force comes directly from the aneroid, while the down-force comes from the spring pressure. Because the available down-force is only a fraction of the available up-force, a situation can arise in which there is enough friction to hang up the mechanism in descent but not in ascent.

Sticky Situation

There could be two sources of the troublesome friction. The lesser of the two possibilities is the cast aluminum frame of the instrument itself. Mounted in the frame is the hand staff, to which the pointers are attached, and according to United's specifications, there should be a clearance ("end-shake") at the end of the hand staff of 0.002 to 0.003 inch, in part to account for expansion and contraction of the frame with changes in temperature. The shop owner says he has found several altimeters that displayed end shake of 0.001 inch or less—in one recent case as little as 0.0005 inch. This could cause binding of the steel shaft in cold temperatures, because the aluminum frame shrinks more rapidly than the shaft as the temperature drops. The reduced end shake, he says, is due to "creep"—a change in dimension over time—of the frame casting. Creep is normal, according to metallurgy experts, and is normally less than one percent.

The more significant source of friction is the presence of unwanted molybdenum disulfide in the instrument. The compound, more commonly called Moly-Kote, is a lubricant normally suspended in oil or grease. It can also be used dry for certain applications. Moly-Kote is currently used in United altimeters only around the edge of the top plate (the face of the frame behind the instrument face) to make the motion of the setting knob smoother. This application is not the problem.

But in years past, United used dry Moly-Kote to burnish some holes in the altimeter's top plate assembly. Afterwards, the excess Moly-Kote would be blown out of the mechanism. According to an attorney for United, the practice stopped sometime in the early '80s. "Not because it was a problem," he said, "but because the manufacturing process had improved to the point that burnishing was no longer necessary."

However, Magagnos says he's been finding it trapped in the small amount of oil that is used for lubrication and contaminating the gears in the top plate. He said he has several work orders showing that sticking altimeters were repaired by the removal of the Moly-Kote. Interestingly, the altimeters cited in the work orders were manufactured prior to 1980.

He attributes the problem to the coagulation of the lubricating oil after a number of years in the field. Magagnos contends the problem wouldn't occur if the lubricant were fresh, even if Moly-Kote were present.

When Magagnos examined the helicopter's altimeter last year, he found that the unit tested within specification at room temperature. However, he said that when he chilled it to simulate the weather conditions at the time of the accident, it hung up while descending. The hang-up required considerable vibration of the instrument to free the pointers.

He said that when he disassembled the altimeter, he found zero end shake in the hand staff, plus contamination of certain bearing surfaces by some sort of oil. After reassembly, he was asked to reproduce the problem but found that he couldn't.

Subsequently, the instrument was sent to a lab to determine what kind of oil was contaminating the bearing surfaces. Magagnos said he had to saw the frame apart so that the lab could examine the bearing surface. Therefore, the altimeter is no longer available for examination.

There is some debate as to whether the inside of the altimeter really was cold at the time of the crash. United's lawyers presented evidence that it wasn't—the only source of cold air is the static line, which by definition has little flow through it except during altitude changes.

The "Fix"

Magagnos feels the problems he has found warrant another AD on United altimeters. He claims to have devised a fix and, in fact, has been independently altering altimeters that have gone through his shop. He has modified more than 70 units to date.

His modification involves removal of any unwanted Moly-Kote in the top plate bearing surfaces, and installation of some new parts. One is a steel brace to keep the frame from closing in and binding on the hand staff. Another is a replacement for the slotted link. Magagnos maintains that if the pin that normally rides at one end of the slot were instead put in a close-tolerance hole, even if there was a hang-up the additional force provided by the aneroid once the pin traveled to the other side of the hole would free it.

But why have a slotted link in the first place, if the pin is always held against one end of the slot? The original reason for having a slot is somewhat obscure, but its current use is well-defined. If spring pressure is removed, the slot allows the altimeter gear train to be checked for proper operation without having to disassemble it, thus simplifying manufacture and assembly. It also supposedly reduces stress on the jeweled bearings by allowing the pin to ride up in it during high G loadings.

Altercation

United got wind of Magagnos's alterations last year and notified the FAA that he was making unauthorized modifications to its altimeters. After examining the alteration, FAA initially said there was no problem with it.

However, the FAA sent a letter forbidding installation of the reinforcing strut or the replacement of the slotted link, since they are modifications requiring special authorization. Magagnos told us he is continuing to modify the instruments despite the letter. He feels he is justified in doing so by a regulation (FAR 145.51) which states that a repair station may make "alterations" of an instrument it is rated to repair.

The agency agreed that altimeters can stick after considerable time in service but maintained that any hang-up would be immediately recognized by the pilot. Magagnos countered that if the hang-up only occurs in descent, and it first shows up during an IFR flight, it could well trap the pilot.

Further, the FAA said that, given friction sufficient to hang up the gear train, even a two-hole link would give an unacceptably high error due to normal clearances around the pin. Magagnos countered with a recommendation of a hole clearance dramatically smaller than that allowed in Tokyo Aircraft's drawings.

The FAA also tried, unsuccessfully, to induce friction by applying excessive Moly-Kote to the bearing surfaces. Magagnos replied that they used fresh lubricant, while the lubricant he says has been causing the problems has been sitting in the instrument for years.

One in a Thousand

The FAA said a review of service difficulty reports (SDRs) for a nine-year period turned up a total of 72 problems with United altimeters, of which only five were sticking. The agency also noted that no other repair shops have reported the discrepancies cited by Magagnos.

Even Magagnos admits that the alleged problem turns up only rarely. "It might only happen in one out of a thousand instruments. We do go through a lot of instruments here," he said. He's seen the problem in only six or eight altimeters over the course of eight or ten years. How does he account for the dearth of similar reports by other shops? First, he feels many altimeter troubles aren't being reported, saying that many shops don't bother filling out the reports.

Second, he says that shops may not be seeing the problem

because of the way they inspect altimeters. There's no requirement for overhaul of aircraft instruments, so the only time an altimeter goes into a shop is when it's broken or needs to be recertified. The recertification specifications in FAR Part 43 only spell out tolerances in ascent. There's no requirement to test the altimeter in descent, and Magagnos says many shops let it unwind very quickly after doing the ascent test. The hang-up friction he's found often shows up only during slow descents.

Cause for Vigilance

It remains to be seen whether Magagnos's allegations constitute a true and widespread safety problem, or whether he is just whistling in the dark. He has presented evidence that at least two accidents have occurred after pilots were led by their altimeters to believe they were higher than they actually were.

While the evidence has convinced juries, it has yet to sway the powers-that-be in aviation. But the story is still unfolding. All of the parties involved in the controversy agree that the alleged problem is so uncommon that the chances of any given altimeter hanging up are quite small. Nevertheless, if Magagnos' scenario is correct, the first time a pilot sees such a hang-up might also be his last. Meanwhile, pilots would be well-advised to check their altimeters carefully during slow descents, *before* they get into IMC where a hang-up might prove fatal. *[Editor's note: Since this article was published in* Aviation Safety, *Magagnos has acquired a TSO to modify altimeters.]*

Instruments Can Indeed Be Liars

Transport category aircraft—airliners, if you will—are equipped with warning devices to let the flight crew know that pitot heat has not been turned on or is not operating. This requirement came about as the result of a Boeing 727 crash several years ago in which the pilot noticed an unscheduled increase in airspeed; he increased the pitch attitude and reduced power in an attempt to control it, and you wouldn't have to be a mental giant to figure out what happened—the trijet stalled, spun, and reached the ground before control could be regained.

"Pitot heat...On" has been on the before-takeoff checklists of most large airplanes for years. Flight at high altitudes—even in the teens—exposes airspeed sensing systems to temperatures well below freezing, and penetration of visible moisture just about guarantees that some ice will collect in the pitot tube. When this happens, the airspeed indicator does some strange things, not the least of which is provide erroneous indications. Here's the story of a Cessna 210 accident in which pitot system icing became a strong candidate for a contributing factor, if not "the probable cause."

The Centurion broke up in flight during an ILS approach to Runway 34L at Salt Lake City International Airport, at the end of a flight from LaVerne, California. Investigators said the pilot and his passengers, all adults, were on the way to a skiing vacation and had not only their normal luggage, but also had six pairs of skis and ski boots aboard. On top of everything else, there was the possibility of an aft center of gravity and/or over-gross weight condition.

The pilot was instrument-rated, had 2,315 total hours, including 1,470 in the Centurion. His wife, seated in the right front seat, had obtained a private license in case she might ever need to take over for her husband, although it was not known whether she performed any piloting duties on the accident flight,.

The 210 took off in VFR conditions on a VFR flight plan, but Salt Lake City—field elevation 4,226 feet—was lying under a fog bank due to a temperature inversion condition. The top of the fog bank was around 5,800 to 6,500 feet, according to various pilot reports.

Controllers vectored the pilot onto the localizer while he was more than 14 miles from the field. A Boeing 737 was in the approach lineup behind the Cessna, and the controller queried the pilot as to the fastest speed he could maintain on the approach. The pilot stated he could maintain 160 knots. Communications were otherwise routine and there was no hint of trouble from the Cessna right up until radar and radio contact were lost.

The wreckage was found on the localizer course, about 1-1/2 miles inside the outer marker. The plane's horizontal stabilizers and elevators, the right wing and a portion of the left wing had separated and were found generally at the same site, about half a mile from the fuselage. The clustering of the separated parts was considered as evidence that they had broken off at nearly the same instant.

The NTSB investigator said the right stabilizer had the appearance of a torsional load, as though it had been "wrung out like a dishrag." The left stabilizer failed in overload. The right wing separated at a point about two feet outboard of the root; the left wing broke at the aileron-flap juncture. The wings failed in negative overload, the investigator said, possibly because the tail failure occurred first and resulted in a sudden pitch-over.

Preliminary weather information indicated that winds at 9,000 feet were from 250 degrees at 11 knots, and would have had little effect on the plane's groundspeed on final. Radar information shows that when the pilot stated he could maintain 160 knots, the airplane actually had a groundspeed of about 180 knots. Subsequently, it decreased to about 175, then increased until as it passed the outer marker, the Centurion was doing 203 knots over the ground. At the last radar response, the plane was at 6,400 feet at 199 knots groundspeed. The investigator noted that the rate of descent stayed in the range of about 400 to 600 feet per minute during most of the approach, but increased dramatically to 10,000 fpm as the plane presumably entered the fog layer.

The landing gear and flaps were not extended, even though a pilot would probably slow down and extend at least the landing gear on passing the outer marker.

The investigator noted that the cloud conditions were conducive to icing, and the possibility of pitot tube icing loomed large. Were the pitot tube to become obstructed, the airspeed indicator would respond somewhat like an altimeter on the descent, and pilot action to increase airspeed by lowering the nose would actually have the effect of helping to decrease the value shown on the airspeed indicator. A little cross-checking and troubleshooting would have gone a long way.

Knowing Your Equipment

Automatic radial centering (ARC) is a labor-saving feature of Cessna navigation radios that has been touted in recent years. Available as an option on the 300 series radios and standard on the 400 series radios, ARC is an attempt to simplify the task of getting radial information from a VOR head.

Most general aviation aircraft have either a mast-type pitot, left, or a tube-type, at right. The mast type usually contains the static port is well, so the static system is somewhat protected against icing.

Cessna's ARC Division (the company) designed ARC (the feature) with a knob in place of the normal OBS which works as follows: With the knob in its middle in-out position, the function is the same as a normal OBS. When the knob is pressed in momentarily, the heading card automatically rotates as necessary to center the CDI needle on the current radial, and the set displays a TO indication with the knob set to the middle position.

When the knob is pulled all the way out, the heading card automatically rotates to center the needle on the current radial, and presents a FROM indication—but it does not revert to the normal position. Unless the pilot pushes the knob back in, it will stay out, and continue to keep the needle centered while turning the card as necessary.

Perhaps you can already imagine how this last nuance could become a trap for a pilot, especially if combined with other problems in navigation; and now it has figured prominently in an accident.

Two experienced pilots—one an ATP with 7,100 hours, the other a commercial pilot with 1,377 hours—undertook a half-hour IFR flight from Columbia County Airport, Hudson, New York to Saratoga Springs Airport in a Cessna 421C. The trip was to position the aircraft to pick up passengers for a Part 135 flight. The ATP acted as pilot in command, but rode in the right seat.

Weather at the time of their arrival at Saratoga Springs is only partially reflected by the sequence at Albany, the nearest official reporting station; measured 500 feet broken, 4,000 overcast, visibility two miles in heavy rain and fog, temperature 60, dewpoint 59, winds 150 at 8. There were low clouds in the Saratoga Springs area, and there was extremely heavy turbulence in the clouds.

Nonetheless, the flight proceeded without incident past the Albany VOR to the GILLA intersection, which is the initial and the final approach fix for the intended VOR approach to Saratoga. GILLA is defined by the 355-degree radial of the Albany VOR (ALB) and the 293-degree radial of the Cambridge VOR (CAM).

They intercepted the CAM 293 radial and flew outbound at 3,100 feet, as depicted on the approach chart. The Albany Approach controller cleared them for the approach and informed them that radar service was terminated.

After a minute and 15 seconds outbound it was time for the procedure turn to the left. At this point, their color radar showed a large patch of bad weather in the procedure turn area specified on the approach chart. It was already bumpy enough where they were, so they decided on an alternative; they would execute a right turn to get on the final approach course. Some IFR pilots may consider the depicted procedure turn as a strictly mandatory maneuver, but there are several options to execute a course reversal, as long as you stay within protected airspace. The depicted procedure turn provides terrain avoidance and pilot orientation.

The two pilots were working together on navigation and decision-making. As they continued around the turn, one pilot set the HSI course selector to the inbound value, but only approximately, figuring he would "fine-tune" the inbound course (113 degrees to CAM) in a few moments.

At this point, there was a discrepancy in the way the pilots interpreted the DME. One told investigators he thought it was tuned to ALB and read approximately 19 miles; the other thought it was tuned to CAM and read about 29 miles. Either way, each pilot could conclude that they were nearing GILLA.

The No. 2 nav head (with the ARC feature) was tuned to

ALB, and showed a centered needle, and the HSI course indicator began to move from right to left, showing them nearing the inbound course. The pilots concluded that they were about to pass a little north of GILLA, and needed only to turn a little more to the southeast.

At 1,900 feet in the descent, they broke out of the clouds, and realized they were very close to the treetops—so close, in fact, that by the time they added full power and began a climb, the airplane actually hit some of the branches. Simultaneously, the Albany Approach controller got an automated altitude alert for the 421, and he radioed a warning to the pilots. They responded that they were already executing a missed approach, and told the controller what had happened.

The pilots turned down vectors for another approach at Saratoga Springs, wisely deciding to go to nearby Schenectady County Airport, a fully equipped facility. They executed an NDB approach to minimums, broke out, saw the field and landed. However, the final event iced the cake; the nosegear, which had been damaged from hitting the trees, collapsed during the rollout.

Later, the pilots told investigators what they had discovered as they retuned their radios while climbing away from the trees. First, the HSI course selector was not precisely on the inbound radial, and therefore the movement of the CDI was not an accurate indication of their position relative to the inbound course. Second, the DME actually was tuned to ALB and read about 26 miles. This, if interpreted correctly, would put them about six miles north of GILLA; but if *mis*interpreted, it could easily convince them they were at the intersection.

The pilots also found the course-setting knob in the full out position on the No. 2 VOR. The ARC feature would therefore keep the needle centered regardless of airplane position, and would "lie" about their location relative to GILLA.

The NTSB investigator said that he was glad the pilots had survived, but gladder still because "if they had crashed and died at the place where they hit the trees, we would never have guessed what decisions led them to be in that position." *Aviation Safety* is glad too, for the lesson it provides other pilots.

Radio Failure and the Law

Part 91 of the FARs requires that an airplane operating under the instrument flight rules be equipped with "navigational equipment appropriate to the ground facilities to be used." While this may seem like common sense, there are situations in which inoperative nav gear need not stop a flight—pilots are expected to "make do" if something fails en route.

Strictly interpreted, the rule can be construed to require that if a particular kind of navaid is to be used, the airplane must be equipped with the appropriate receivers, or the intended flight is not "legal." And even more strictly, if the equipment cannot receive the navaids on both the intended approach and the missed approach procedure, the pilot may be considered a rule-breaker if he executes the approach. Pilots flying under Parts 135 and 121 are held even more tightly to these standards.

All this should be obvious; even if there were no rule, a pilot wouldn't try an ILS without an operating glideslope receiver (chances are, he could land on the same runway with a localizer approach, anyway), nor would he attempt a VOR approach without at least one working VOR receiver.

But what about the ADF? Aren't there many approaches where an ADF is only a useful addition, not an absolute necessity? What if the missed approach procedure is totally dependent on an NDB? What about a marker beacon receiver—can canny navigation obviate its use? Issues like these—plus a navigational blunder and some maintenance questions—are being analyzed in the wake of a Piper Navajo crash that took the life of an experienced courier pilot.

The crash occurred in the wee hours of a November morning when the Navajo plowed into a mountain near Asheville, South Carolina. The pilot held an ATP with 3,000 hours of which 2,000 were multi-engine, 1,200 at night, 427 actual instruments, 900 in Navajos. For several months, the pilot had made a regular run from Greer, South Carolina, to Asheville, North Carolina, to Columbus, Ohio, returning by the same route.

On the night of the accident, the pilot left Greer as usual shortly after 9 p.m., stopped at Asheville, and completed his trip to Columbus. At 2:30 a.m., he was airborne on the return portion of the flight, which was routine until low visibility forced a missed approach at Ashville. The pilot

Despite their expense, aircraft radios fail from time to time. The FARs provide for this eventuality but a better plan is to have a hand-held transceiver as a back-up.

elected to continue to Greer, and upon receiving clearance direct to the Spartanburg VOR then direct to Greer, the pilot told the controller he would "intercept the 235 radial." This inferred that the pilot would proceed generally south and east, intercept the 235-degree radial of the Spartanburg VOR, then fly it outbound to Greer.

It should be noted that the approach controller did not have reliable radar coverage below 7,800 feet in the area, thus navigation was generally up to the pilot. Also, the controller characterized the Navajo's transponder as "weak," and the full beacon target was observed for only a short period of time after it climbed on the missed approach and headed southeast.

The controller needed to verify the plane's position, so he asked the pilot to estimate his arrival time at Greer. The pilot did not respond directly to this request, but said "I'm on the 235-degree radial now." The pilot was cleared for the ILS Runway 3 approach, reported going through 4,400 feet on his descent, and that was the last the controller heard.

Four of the six popular handheld transceivers also have VOR capability. In emergency conditions, an approach could be flown with one.

The wreckage was found at the 3,200-foot level of a 3,700-foot mountain, 24 miles out on the 236-degree radial of the Sugarloaf Mountain VOR. Because of near-total destruction of the plane and its instruments, very little could be learned from the wreckage, and much of the investigation was necessarily inferential. According to maintenance records, the plane had been squawked for an inoperative right vacuum pump, and a left engine that surged in climb, but the NTSB investigator said these items may not have had a bearing on the crash.

Most significantly, the investigator found that the Navajo's ADF receiver had been inoperative since the previous May. Further, the light that illuminated the No. 2 OBS was inoperative. The nav radios themselves were of a digital variety which allows display of frequency or radial information, but not both at once. The marker beacon receiver was inoperative in both aural and visual modes. The pilot may have been conducting NDB approaches without an ADF, and he flew this flight without a marker beacon receiver.

This introduces a possible scenario for a pilot with such equipment gaps. If he were conducting an ILS at Asheville without an ADF, he would not be able to pick up the Broad River NDB, which is the final approach fix for the localizer version of the approach. But (assuming his No. 1 nav is on the ILS), he could tune his No. 2 nav to Sugarloaf Mountain VOR, whose 234-degree radial crosses the localizer at the NDB. When the needle centered, he could announce his position at the fix. Since this also is the point at which glideslope interception can be expected, there is little consequence to the lack of an ADF.

But there is a problem with the missed approach procedure, which calls for a climb to the KEANS LOM and hold. This is theoretically impossible without an ADF, although an experienced pilot might be able to estimate the LOM location and imitate a holding pattern. But, especially late at night when controllers aren't busy, a pilot might also simply climb straight ahead, and count on being given a clearance to a known navaid before it became necessary to hold.

If a pilot making such a climbout were now given a routing to Greer via the Spartanburg VOR, he might turn to the southeast, get out the approach plate for Greer and notice that he would be expected to fly the 235-degree radial outbound to the Greer LOM, the initial and final approach fix. Confirmation of his arrival at this fix would present essentially the same task as during the Ashville approach, although some help would come from a DME reading as he flew out the radial.

If the pilot now noticed that his No. 2 nav was already set on almost precisely the needed radial, he might simply adjust that and forget that the radio was tuned to the wrong VOR. This is the kind of blunder a pilot can make all too easily at the end of a late-night flight on a very familiar route—complacency strikes again.

Circumstantial evidence, true; but it should be enough to give any IFR pilot some food for thought, especially if he is now flying with some radio equipment inoperative.

I Can't Talk—I Can't Hear—I Can't Navigate!

Contemporary aircraft radio equipment is rather reliable. We complain liberally about the quality of some of the sets, and trouble shows up on occasion, but for the most part we are able to communicate some way or another nearly all the time—the incidence of complete communications failure is a rare event. And when it does happen, there are rules and procedures guaranteed to get you out of trouble. The rules are written in considerable detail, but in general, you're expected to fly at the last-assigned altitude, along the last-assigned route, and execute the approach procedure of your choice when you reach the destination airport. If everything on board quits and you can't navigate, revert to the magnetic compass and head for the nearest VFR conditions.

When the Electrons Stop Flowing

Some multi-engine pilots insist that the real reason they prefer to fly behind two engines (or three, or four or more) is system redundancy, especially when operating in IMC. A twin provides two of everything—engines, vacuum pumps, alternators or generators, fuel pumps, and so forth—and when navigation and communications are completely dependent on electrical power, it's awfully nice to have a built-in standby source. There has been a flurry of interest in a second electrical source for single-engine airplanes; the new designs (and there are but a handful since single-engine development has all but ceased in the U.S.) often feature standby generators as standard equipment, but most pilots must accept the risk of electrical problems when a component of the system fails.

The certification rules speak to the need for system-condition monitors, mostly commonly represented by either an ammeter or a load meter, and often combined into one instrument. It's important for IFR pilots to understand what they're seeing, because a failing electrical "pump" often takes the system downhill in a manner that can only be described as insidious. In the absence of generator/alternator failure warning lights, the only early indication will be an increased drain on the battery, and by the time the signs become unmistakable—lights dimming, nav needles flickering, radio communications getting progressively weaker—the amount of electrical power remaining in the battery may not be enough to complete the flight.

Here's the report of a mishap that involved the electrical system of a Cessna 210 flying in IFR conditions. The problem was well annunciated, the pilot did everything properly, right up to the very end of the episode, when a reversion to normal habit patterns in an abnormal situation resulted in a bent airplane. Very fortunately, there was no damage to the pilot or his two passengers.

The pilot had logged more than 2,000 hours in the air, including 900 in the Centurion and 583 hours on instruments. This trip had begun in Florida and was bound for Danbury, Connecticut, when the voltage warning indicator came on and the pilot made a stop in Richmond, Virginia, to attend to the problem.

The mechanic found problems enough in the electrical system to recommend replacement of the alternator and voltage regulator, so new units were installed and thoroughly ground-checked. The pilot set out on the remainder of the flight, a two-hour leg to Danbury.

All of the Cessna's avionics—including weather radar—were in use, and in the vicinity of New York's Kennedy International Airport at 4,000 feet, light icing required the occasional use of the de-icing equipment. At this point, the pilot noticed a flickering of the cabin lights, and although the low-voltage warning light had not illuminated, the ammeter showed a full discharge, indicating that the plane was running solely on battery power.

The 100-amp alternator circuit breaker tripped promptly after one reset, and would not reset again. The pilot immediately began turning off all non-essential equipment and requested vectors to Westchester County Airport, the nearest reasonable destination, where the weather report indicated a 600-foot broken ceiling.

Arriving overhead, the pilot reported the airport in sight, and when the controllers issued vectors to line him up for an ILS Runway 16 approach, the pilot requested a visual approach. Of course the airport was officially IFR, and the controller had no choice but to clear the pilot for a full-blown instrument approach. Choices changed when the pilot stated that he did not think the battery would allow an IFR approach (his primary nav radio had already begun to flicker), and the controller therefore cleared the pilot for a contact approach.

The pilot proceeded to land on Runway 11, almost without incident. As he told investigators, "Without thinking, I lowered the landing gear and flaps using the electrical switches. I do not recall seeing a green (landing gear) light, or even looking for one." The flaps reached about 20 degrees extension and the gear extended only partially before battery power was apparently exhausted. The plane slid off the runway and stopped.

Among their preliminary findings, investigators discovered that the screws connecting the wires to the alternator circuit breaker were loose, and the connectors themselves were discolored, possibly due to overheating. The pilot told investigators that the plane could have been equipped with a standby generator or alternator. "This plane originally had one, but it was removed in order to comply with the Airworthiness Directive requiring a second vacuum pump," the pilot stated.

A Question of Tolerances

A Piper Warrior crashed into Maiden Peak, southeast of Port Angeles, Washington, in 1981, killing the two occupants and leaving only a tiny dent in the mountain. But the implications of what investigators and FAA personnel learned in the aftermath of the crash may be felt throughout the nation.

In a nutshell, the Warrior's crash caused FAA to consider this question: Even though an instrument airway meets all applicable standards and passes tests conducted by FAA flight check aircraft using sophisticated instrumentation, can a general aviation aircraft, with its relatively low-cost VOR receivers and typical low-quality antenna installation, actually fly safely down that airway?

While there isn't enough information to provide an absolute answer to the question, the qualified answer is a rather startling "no." There is a reasonable possibility that a typical general aviation light aircraft, even though equipped to all applicable standards for IFR flight, may not be fit to fly certain IFR airways safely.

The fault probably does not lie with the avionics, but with the installation, especially the low-cost antennas which bedeck typical small aircraft, the placement of those antennas, and perhaps the maintenance of the antenna and radios as a system. More expensive aircraft, with more expensive antenna installations, are much less likely to suffer from the problems, if at all.

It should be pointed out that certain circumstances make the Port Angeles case extremely unusual, and that most Victor airways are as safe as pilots have always presumed them to be. But there is the potential that certain airways—especially when they stretch long distances away from the nearest VOR station, and when they lie in close proximity to high terrain—may be unsafe for small aircraft to attempt to follow, even though the airways themselves are certified, properly established and monitored.

The crash resulted in the most obvious and simplistic response by the FAA; it raised the minimum en route altitude (MEA) of the airway in question. While this cures the specific problem, it does nothing to address the question of how many other airway segments might pose similar problems.

An alternative would be to create a standard for the whole airplane, with antennas and radios, so that any IFR-certified aircraft is assured the ability to fly safely along any IFR airway. On the other hand, airways flight inspections could be conducted with equipment that imitates the lowest-quality general aviation installation, rather than with the top-quality equipment now used. These were among the suggestions raised by FAA regional officials to their superiors in Washington shortly after the crash.

Northwest Region officials had conducted a test in which they rented four typical general aviation aircraft and tried to fly the airway as the Warrior pilot had done. Three of the four aircraft could not stay within the confines of the airway, even though their nav receivers were checked and working within specifications before the flights, and even though the VOR signal was checked and within all agency specifications.

The Northwest Region said to Washington, "We recommend that action be initiated to determine if typical general aviation aircraft navigation receivers and antenna systems provide satisfactory guidance information to safely use our present airway system. If further evaluation substantiates our results, we believe airworthiness standards for these receivers and antennas may need to be revised, or our flight inspection criteria modified to consider performance of these systems."

Nearly two years later, when *Aviation Safety* asked what FAA headquarters had done in response, the agency gave contradictory and confusing answers, and was unable to give specifics on any action taken. Even experts in the field appear to be baffled by the questions raised, so the lesson for pilots and aircraft owners is that performance of antenna systems is not to be assumed, and must be constantly monitored.

Beyond the large issue of the fitness of general aviation aircraft

The FAA regularly flight checks all airways for accuracy and signal strength but equipment used in flight check aircraft is far more sensitive than that found in the average general aviation aircraft.

to negotiate mountain airways, the Port Angeles crash also poses lessons that may be valuable for all IFR-rated pilots, and even some VFR-only airmen as well.

Simplicity of Results

The Port Angeles crash can be summarized with extreme simplicity: A pilot departed IFR, attempted to follow a Victor airway at or above the MEA, and struck a mountain off to the right of the airway. Later investigation showed that some—perhaps many—light aircraft with low-cost radio and antenna installations might also have been led into the mountains in the same way.

However, there are many facets of the accident which lend themselves to study for their own sake, and the simple result—hitting the mountain—was not so simply arrived at, nor can the absolute truth ever be known.

John H. Ohm, a young flight instructor, had been climbing the ladder of pilot ratings; a week earlier, he had added the instrument instructor rating to his CFI certificate. He had a total of 695 hours, of which 552 were in Cherokees. He had 31.7 hours of night flying. He had logged only 7.2 hours of actual instrument time (4.2 of it as pilot in command), but he also had some 62 hours of simulated instrument time (17 of it as PIC). Although Ohm was current and qualified to fly IFR under Part 91 and instruct students, he was not qualified to fly a charter under Part 135.

On January 28, 1981, Ohm was at his FBO lounge at Bellevue Airport (a Seattle airport not far from Boeing Field) when a potential customer came in and said that he wanted to hire an aircraft to go to Port Angeles. According to the statement of one pilot on hand at the time, the instructors (including Ohm) at first told him there was no airplane and pilot available; but when the customer expressed disappointment, Ohm volunteered to move other students in his day's schedule to allow time for the trip. "I can take you out to Port Angeles and I will charge you like a student on a cross-country," said Ohm, and the customer agreed to this arrangement.

Ohm had called for weather and filed his flight plan when the FBO's chief pilot showed up and learning what was intended, told them the trip couldn't be made unless the customer were a pilot or student pilot. Ohm advanced another suggestion; "What if I rent the plane and he just pays for the rental cost?" This plan was accepted.

In essence, Ohm would perform the flying free of charge, conducting a Part 91 flight with his passenger paying only for the airplane. He would be taking advantage of Part 61.118(b), which allows anyone with a private certificate or higher to share the aircraft operating expense with his passengers. "I need to build up time anyway, and it's about the only time I get to fly anywhere," Ohm told the pilot-witness.

We must note that Ohm's understanding of the rules could be disputed. FAR 61.118(b) is usually construed to mean that first, the intended flight must be primarily for the benefit of the pilot (not to transport the passenger) and second, the sharing of expenses must be in the ratio of the number of people aboard (i.e., a pilot and one passenger each pay 50 percent).

ARRIVAL AT PORT ANGELES

The NTSB records do not contain detailed information about the flight to Port Angeles, but it can be assumed that it was an IFR flight that arrived there around 2 p.m. The customer went off to keep his appointment, and Ohm visited with a local pilot, who told the NTSB that Ohm talked about his ILS approach into the airport. "He had flown into some light ice and forgotten to turn on his pitot heat. He noticed this on his instruments and turned on the heat, but in the process he was high on the glideslope and declared a missed approach. He made the second approach with no difficulty." This witness was present as Ohm phoned the FSS for the return-trip weather, then again to file the flight plan. "He had the en route chart in hand and knew the route he was to take. His flight plan was not rushed, nor did he seem confused on any of his procedures. I did not see the SID chart in his planning or papers, so I assume he was departing as written on the back of the approach chart," the witness said.

The witness said he watched pilot and passenger board the Warrior, taxi out and hold for 10 minutes, then "they took off on Runway 8 straight-out and disappeared into the clouds." It was about 4:46 p.m., and getting dark.

WHICH IFR DEPARTURE?

What occurred between the takeoff and the subsequent impact on Maiden Peak is open to some conjecture, since there is no independent means (such as radar data) to determine where the Warrior was, and since several options were available to the pilot. An unresolved ambiguity in the clearance left Ohm at liberty to fly the published instrument departure procedure contained in the approach-chart book or continue straight ahead in a climb to join the airway, Victor 4. A comparison of the content and timing of the ATC transcript with the performance of the Warrior seems to develop a solid correlation indicating Ohm used an on-course climb to join Victor 4.

Ohm's initial clearance to 5,000 feet was later amended to 6,000. Whatever procedure he used, it is clear that he got to 6,000 feet and was attempting to follow the airway when

he crashed into Maiden Peak at the 6,000-foot level.
Here is the transcript of communications:

1647:09 Pilot: Seattle Center, Cherokee 3062D with you, climbing out of 1,700 for 5,000.

1647:16 Controller: Cherokee 3062D, roger climbing to 5,000. Report level and, uh, verify you are squawking 4567.

1647:24 Pilot: Roger, 4567, 62D. Verify 5,000.

1647:28 Controller: Roger, climbing to 5,000 and report level.

1647:32 Pilot: 62D, report level 5,000 and, ah, report level at 5,000.

1647:37 Controller: Roger.

1649:06 Controller: Cherokee 62D, say altitude leaving now.

1649:50 Pilot: 3,600, 62D.

1649:52 Controller: Roger.

1651:31 Controller: Cherokee 3062D, climb, maintain 6,000 and report level. Say altitude leaving now.

1651:36 Pilot: 62D is out of 4,800 climbing to 6,000.

1651:41 Controller: 'Kay, sir, report level six. I've got other traffic I'm trying to clear out now.

1651:44 Pilot: Okay, 62D.

1653:09 Controller: Cherokee 62D, altitude (unintelligible) leaving now.

1653:22 Pilot: Five thousand, nine hundred, 62D.

1653:25 Controller: Roger.

1653:50 Pilot: Seattle Center, 62D level 6,000.

1653:53 Controller: 62D, roger.

1653:54 Pilot: (Unintelligible) you get a current altimeter setting, 62D?

1653:57 Controller: 62D, Port Angeles altimeter is, ah, 29.62.

1654:03   Pilot: Thank you very much.

1658:20   Controller: 3062D, Seattle. Report the JAWBN intersection. Be advised, negative on your transponder yet.

1658:33   Controller: 3062D, Seattle.

1658:50   Controller: 3062D, Seattle.

1658:58   Controller: 3062D, Seattle.

Wreckage Evidence

The Warrior had never been observed on radar, but within minutes after the crash, there was a report of an emergency locator transmitter signal in the area of Maiden Peak. The following day, searchers found the wreckage on the north slope of the mountain. Both occupants had died on impact.

No evidence was found that would suggest pre-impact malfunction of the airplane or engine. The aircraft was probably in level flight when it struck the mountain, skidded up a 55-degree slope for 60 feet and came to rest on a bearing of 175 degrees from initial impact. Because of the terrain, this was not a reliable indication of its final pre-impact heading. Time of impact was reckoned at 1655, which is when the pilot's watch stopped (the aircraft clock stopped at 1709). Among other panel findings were the following:

Comm 1: Frequency 125.10 MHz, on, volume one-fourth.

Comm 2: Frequency 122.8, volume one-fourth.

Nav 1: Frequency 114.20, ident position, volume three-fourths, OBS 240 degrees.

Nav 2: Frequency 111.20 or 112.20, volume three-fourths, OBS 083 or 084, jammed.

ADF: Frequency 260 kHz, needle 145 degrees, volume one-fourth.

Altimeter: 29.61.

Directional gyro: 115 degrees.

It should be noted that none of these findings is absolute, since instrument needles and even frequency dials have been known to move during a crash sequence. However, all are quite consistent with local navaids.

While there may be many possible interpretations of the findings, we suggest the following possible scenario: The pilot was
talking to Seattle Center on Comm 1, and had monitored Port
Angeles Unicom on Comm 2 while departing. He had tuned the
ADF to the ELWHA LOM. We suspect there was no valid DME
signal in the area of flight, and that this would not surprise the pilot.
Nav 2 was tuned to Tatoosh VOR as a backup to Nav 1 at takeoff.
In all likelihood, the pilot had also used it as a backup during the
previous ILS approach, with the OBS at 083 as the final approach
course. Now tuned to TOU, it would be a reference during the
departure, but perhaps not a good one.

Nav 1 had probably been tuned to 114.20 just before impact as
the pilot, becoming concerned about his navigation, sought to find
a cross-reference from a closer navaid with an unobstructed signal.
The frequency is that of Paine VOR, across Puget Sound to the east.

We believe the pilot had just confirmed Paine, was centering the
needle and about to discover he was off course when the Warrior hit
the mountain. This is bitter irony; had he recognized his position a
little earlier, the pilot might have escaped death. A second irony is
that the controller did not mention the lack of a transponder return—a warning to the pilot that there was no independent means
to help him navigate—until the pilot most likely was already dead.

Where was Nav 1 tuned before takeoff? We believe it was on the
TOU frequency, with the OBS on 080 degrees, the definition of
Victor 4.

Airways Flaw?

Investigators looked at the findings and inferred essentially the
same things. Regardless of how he got to the airway, the pilot had
attempted to fly Victor 4, but wound up eight miles to the right of
course. Investigators noted the two nav receivers had been checked
just two days prior to the crash, with logged errors of one degree and
zero degrees recorded. There was no evidence to suspect out-of-
limits receivers.

At first there seemed to be no reason to question the airway and
navaids. Victor 4 had been reviewed for terrain clearance and the
TOU signal had been flight-checked at MEA before the airway was
commissioned. A day after the crash, FAA sent out a flight-check
aircraft and confirmed that the signal met all applicable standards.

But just a week after the crash, in response to a pilot complaint that the TOU signal was erroneous in the area east of Port Angeles, the FAA performed another flight check and got the same results; the investigators began to suspect there was a difference between an FAA flight-check aircraft and a "real" general aviation aircraft. The FAA uses sophisticated jets with numerous navigational systems and a good autopilot to integrate the guidance information. The flight-check aircraft, equipped with the very best navigational antennas and the most sensitive devices to detect and record signals, flies straight down the center of the airway, rather than pursuing any erroneous needles.

The typical general aviation aircraft is remarkably different. Its radios may or may not be TSO'd, they may have inherent, allowable errors, and they are far less sensitive than the flight-check equipment. But more important, the lightplane's antennas are at best nowhere near flight-check quality, and there are many reasons why a lightplane's antennas might not be working perfectly (design, construction, installation, lack of maintenance).

It should be noted that a "sum of allowable errors" fails to explain the discrepancy. Depending on how it is checked, a VOR receiver may be inaccurate by as much as four degrees. A VOR signal may have an alignment error of as much as half a degree, and it's signal may be allowed to exhibit scalloping of as much as two degrees. In the worst case—all errors stacked up in the same direction—the pilot could get 9.5 miles off course. However, the VOT log entry, checks of the actual Tatoosh signal and other evidence produce an error of only about 6.5 miles at most, and more likely only 2.0 miles—not enough to explain an eight-mile error.

The investigators settled on a practical test. They obtained a Piper Seminole with factory equipped radios like the Warrior had (two King KX-170Bs) and essentially the same antenna system. They flew the Seminole westbound on Victor 4 from JAWBN to Port Angeles at 5,000 feet, and received a good, valid signal from Tatoosh.

Then they flew eastbound from Port Angeles, using the TOU 080-degree radial; their finding: "A fly-right indication was noted on the CDI beginning about 49 nautical miles east of Tatoosh (i.e., over Port Angeles). Course adjustments to the right were required to keep the needle centered. Adjustments to the fly-right indication continued to divert the flight south of course until inadequate clearance from terrain was foreseeable. The flight was terminated 57 nautical miles southeast of the Tatoosh VORTAC."

They tried the same thing at 6,000 feet. "Initial course deviation began to occur approximately 51 nm east of the facility (TOU) and the flight was terminated in mountainous terrain at 59 nm." Next, they tried it at 6,000 feet with the OBS on 083 degrees. "The flight was terminated within close proximity to Maiden Peak."

The investigators performed a VOT check on the Seminole's radios prior to the flights and recorded +1 and -1 degree errors. They found nothing to make the radios suspect, and *Aviation Safety* found nothing to impugn the accuracy of these or any other general aviation nav receivers. The issue lies with the antenna and installation.

The NTSB eventually pronounced a most unusual probable cause; "Airways facilities: VORTAC. Remarks: Airway unsuitable for navigation at assigned altitude. Reduced signal strength beyond 40 nm."

FAA Action

Meanwhile, officials in the airways facilities section of FAA's Northwest Region had been informed of the findings and treated them very seriously. They wanted to find out why the flight-check aircraft could detect nothing wrong with the airway, but light-planes couldn't fly it.

Several typical general aviation aircraft were rented, flown by FAA staff members to see how they performed on Victor 4. One of these was a 1979 Cessna 172 with a "towel bar" antenna and two KX-170B radios. This aircraft consistently flew Victor 4 south of course, whether west- or eastbound. Out at the end of the airway, its radios indicated the JAWBN intersection when the plane was actually 12 miles south.

A 1979 Mooney 231 was also rented for the test. It had a towel bar antenna, two KX-170B radios, and again flew to the right of course eastbound, "placing the airway south of its actual position."

Then the testing moved up a notch; a 1978 Beech 58 Baron, equipped with blade-type nav antennas and two KX-175B radios flew the airway in both directions with no deviations.

An FAA staffer reached the conclusion that the signal strength—at least the strength of the signal reaching the nav radio—was not sufficient for navigation. He said many of the flights involved nav flags that were intermittently "peeking" into their windows. One thing was quite clear in the FAA staffer's recollec-

Fresh from the factory, most singles have cat's whiskers VOR antennas, instead of the blade-type found on more expensive singles and twins.

tion; virtually all of the planes which demonstrated course excursions were equipped with an open-loop dipole type of VOR antenna, commonly nicknamed a "cat's whisker" antenna.

Whisker-type antennas are relatively directional, exhibiting their best reception in the areas fore and aft of the airplane, and worst from off to the side. This tends to make a whisker antenna somewhat more susceptible to VOR signal scalloping effects, in which signals from a VOR may bounce off mountains and reach the antenna from a different angle than the direct signal. The two signals now combine to form an erroneous signal that is sent to the nav radio, which has no ability to select and display the true signal.

In contrast, balanced-loop antennas have virtually the same reception capabilities in all directions, making them superior for navigating with VORs off to the side of the course, and less susceptible to scalloping problems. Balanced-loop antennas may come in the form of "towel bars" (beware, though—some towel bar antennas are electronically no different than whiskers), or in more expensive versions, such as small blade-type antennas.

The Northwest Region testers found not just the type of antenna was of interest, but also its placement. They discovered whisker antennas were sometimes installed in questionable locations on the airplane, such as high on a T-tail, where aft reception might be blocked by the horizontal stabilizer.

They also observed that reasons for signal loss could include not only the antenna, but the parts of the system leading from there to the radio.

For instance, a whisker-type installation often has a device on or near the base of the antenna for impedance-matching to the coaxial cable. The device, known as a balun, may be subject to corrosion and other problems of advancing age. (One expert on such installations said because baluns are often stuck inside the vertical fin with little protection against moisture, he's seen corrosion occur on factory-installed baluns only six months old.) This can cut the strength of the signal before it enters the coax cable.

The coax cable may also deteriorate and crack due to the hot and cold cycles to which an airplane is exposed. This is another source of signal strength reduction.

Virtually all light aircraft with dual VOR radios use just one nav antenna to supply signals to both receivers, making it necessary to have a signal splitter to divert the signal down two coax cables to the radios. The coupler is responsible for a large loss of signal strength.

Finally, at least one expert suspects the press-fit connection between the last piece of coax and the radio itself as a source of signal strength loss.

When all these effects are combined, it is not unreasonable to find the signal strength cut in half between the antenna and the radio, and sometimes lots worse than that.

We were intrigued by what avionics technicians told us about general aviation receivers. When a nav radio is bench-checked for accuracy, the signal typically is at a level of 100 microvolts (uV). But if the radio is checked for the signal strength that gets the nav flag to roll out of the window, this level is generally around 2 uV. In other words, a nav receiver unflags at a much lower signal strength than that of the accuracy check. We are advised that the design of the radio makes it likely that in the region between 2 and 100 uV the receiver is as accurate as at higher levels, but there is no test normally made to confirm it.

In actual practice at normal altitudes, many VOR transmitters deliver a signal strength of perhaps 100 uV to the plane's antenna.

Flatlanders and those who stay within 30 miles of a VOR almost never have a problem.; but the FAA says that some airway segments could see as little as 25 uV and still be certified.

That's the signal the antenna receives, but because of the various losses on the way to the nav radio, the 25 uV may be dropped to 12 uV, or even less. The level may be enough to roll the nav flag off, but certainly not enough to create a reliable indication on the CDI. If the needle were obviously erratic, the pilot would ignore it, but there are some cases in this low-level signal area, particularly where scalloping is a concern, that may create a relatively smoothly yet deviant needle.

Thus it is possible to have an airways VOR signal that meets all the standards being used by aircraft whose antenna systems do not deliver that signal to the nav radios in sufficient strength for safe navigation—and pilots may never be fully aware of the problem. An airplane may be certified for IFR flight and the pilot may or may not be satisfied with the VOR reception he gets, but he may accept this performance as the norm. The radios are not the problem, periodic bench-checks are irrelevant—the bench check assumes a given signal strength going into the radio. Even a VOT check will not reveal the problem, since it is highly unlikely that a pilot would use a VOT facility with a poor signal.

A seasoned IFR pilot might realize that a "safe" nav flag doesn't always mean a valid VOR signal is being received. A veteran mountain pilot who has been over a route VFR will also observe areas of course scalloping and remember not to trust the CDI in those areas. But the combination of a less-experienced pilot, an unfamiliar airplane, wet weather conducive to scalloping, and an airway defined by a distant VOR may create the same problem that put the Warrior on Maiden Peak.

Enormous Implications

The FAA staffers doubtless realized the implications of what they had found. A look around any airport will reveal that the vast majority of general aviation airplanes are equipped with cat's-whisker antennas, cables and other system components closely matching that of the crashed airplane and the ones they tested on Victor 4.

Aviation Safety found that most light aircraft are equipped with a whisker antenna and installation similar to the Warrior's unless a buyer specified balanced-loop antennas. The manufacturers generally offer the higher-cost balanced-loop antennas as optional equipment in more expensive aircraft, and some recommend it where area navigation systems are installed, since the whisker antenna is poor in receiving stations at right angles to the airplane. In stark contrast, Mooney Aircraft has adopted a policy of equipping every aircraft with a balanced-loop antenna.

Although they had "fixed" the local Victor 4 problem by raising the MEA, the FAA realized that there could be other Victor 4s in existence elsewhere in the country, or perhaps about to be created; and they realized that no standard exists to guarantee that Victor 4s won't occur, or that pilots won't try to fly them.

Preventive Measures

Pilots can do several things to protect themselves from being caught by a "Victor 4." Among them:

• Consider installing a balanced-loop antenna system. This will improve regular VOR reception, decrease the likelihood of scalloping signals, and if there has been an Rnav installed recently, vastly improve that unit's usefulness.

• At annual inspection, assure that the mechanics inspect the whisker antenna and its balun, and replace either component if there's any sign of corrosion or deterioration.

• On an aging airplane (10 years old or more), consider replacement of the entire antenna system—all the way to the receivers—at annual inspection time.

• Under IFR conditions, never fly a route in mountainous terrain that has not been flown in VFR conditions with attention to whether valid nav signals were being displayed—particularly if the altitude to be flown is at or near MEA, the airway relies on a single VOR to define it, and the distance is more than 40 nm from that station. In short, never rely on an airway to be totally safe for you and your airplane, just because it's on the chart.

No Firm Conclusion

Three out of four typical light aircraft had been unable to follow the airway (actually, four out of five, if you count the Warrior that crashed). The FAA staffers had demonstrated there was a big problem. Now the big question: Why? Unfortunately, there wasn't (and isn't) a simple answer, because it may be the sum of many smaller problems that can be divided into two areas; something wrong with the signal, and something wrong with the airplane.

The Signal

A first issue that must be addressed is signal scalloping, a situation that occurs when a VOR receiver gets more than one signal. The primary signal arrives direct from the VOR, but in certain circumstances, another signal can arrive at the airplane after having taken an indirect route, such as bouncing off a mountainside. Unfortunately, when the two signals are received at the aircraft antenna, they may blend into a new, erroneous signal. Especially with low-cost antennas, there is no way to discriminate between the direct and the multi-path signal. The receiver gets whatever the antenna has collected, erroneous though it may be.

Where a multi-path condition exists, a VOR signal is said to "scallop." If a pilot were able to use some other means to track a perfect course on the airway and could judge the VOR in relation to this line, he would see the needle move to one side or the other during portions of his flight. If he had to rely on the needle for guidance, his course would "scallop" slightly.

It is important to distinguish the relatively slow, misleading needle movement produced by scalloping from other effects that produce quick and erratic needle movements (e.g., when approaching the station). Most airmen would probably ignore an erratic needle but, unless they had other information, would probably follow a scalloping needle.

Scalloping almost never occurs in perfectly flat land when the airway altitude ensures line-of-sight to the VOR; it generally requires mountains bounce the signal. Ideal conditions include a steep mountain ridge running roughly parallel to the course to be flown, but there is another factor—moisture. Dry and tree-covered mountains may not create as good a mirror for the signal as wet, rocky, or snow-covered mountains.

Which brings up a problem. The FAA does not commission airways without concern for scalloping, and they make careful tests to ensure that scalloping does not exceed two degrees anywhere along an airway. Victor 4 was tested on numerous occasions, and of all the results we've seen, scalloping never exceeded 1.5 degrees—

sometimes, it was zero. The trouble is, flight checks are almost never conducted in the same weather conditions encountered by instrument pilots, and therefore the worst scalloping on any given airway is probably never measured.

Secondly, FAA flight-check aircraft do not imitate general aviation aircraft. The flight-check aircraft has other ways to know where it is (multiple DME and inertial navigation systems) and it flies carefully down the center of the airway while measuring the meanderings of the VOR signal. The general aviation aircraft follows the signal; if it meanders, so does the aircraft.

It is at least theoretically possible for a scalloping signal to get worse the farther off the airway it leads the airplane. However, this would take extremely special terrain geometry. In any event, on at least one test an FAA flight check aircraft sought to imitate the off-airways course of the Warrior, and did not detect significant scalloping until it was already far into the mountains.

Vital Connections

No matter what the antenna, it merely collects the signal, which still has some distance to go before becoming useful to the pilot.

Next after the antenna, usually mounted in the tail or the base of the antenna itself, is an impedance device to match the antenna to the rest of the system. It is called a balance-to-unbalance transformer, or "balun." Some extremely gross field installations may omit the balun, in which case the airplane's radios will perform very poorly. Most installations include the balun—but once it's installed it may never be inspected again.

Lots of airplanes have dual receivers, and it is necessary to send the antenna signal to each of them. So the coax from the antenna leads to a coupler, or "splitter." Out of the splitter come two pieces of coax for the individual nav radios, and sometimes a third one for the glideslope receiver, although a superior installation has a separate glideslope antenna.

It should be noted that the comm antennas for the same radios are entirely separate. An owner may see two comm antennas and think he has total redundancy; actually, due to siting difficulties, there is usually only one VOR nav antenna to service both nav receivers. Finally, a last connection mates the coax to the radio.

This chain of connections is absolutely vital, since every point along the way provides a danger of decreasing the signal strength.

The balun, the coaxial cable, any connectors, the splitter and even the final connection to the radio may all be sources of signal attenuation.

How much can the signal be cut? In a typical lightplane, these devices amount to a halving of signal strength even when the whole system is brand-new and properly installed.

As might be imagined, some of these devices may develop problems that cut the signal even more as the airplane gets older. One expert told us he's seen baluns only six months in service already heavily corroded. Hot-cold cycles can eventually cause the coaxial cable to crack. Connectors may be sited where there is moisture collection. Splitters usually do not suffer problems, but the final connection to the radio may not always stay in good shape as a radio is pulled out for repairs.

Some of these problems are the kind that lead to obvious clues for the pilot; extremely poor radio performance and interference when using the audio to confirm a station identifier. However, other problems can simply cut the strength of the signal reaching the radio, and not give obvious clues. Unless he has reason to fly in an area where signal strength is relatively low (and made a lot lower by his deficient antenna system), he may never notice a problem.

The careful pilot who regularly conducts VOT checks also will not detect a signal strength problem. No pilot logically would conduct a VOT check with a facility that he receives poorly; therefore he never investigates the performance of his radio with a weak signal. There is no FAA standard specifying—nor any requirement for periodic checks of—the performance of the airplane's antenna system in the field.

The Radio

So far, the nav receiver has not been called into question. This is because nearly everyone involved in the Victor 4 investigation believes the radio is not the problem, but rather the signal reaching the radio.

However, there is one aspect of the radio which may have a bearing. If a signal is so weak as to be unsafe and unusable, shouldn't the radio provide a flag? And if the flag goes out of view, shouldn't that mean the signal is safe and usable? We were not encouraged by the answers we got to these questions.

Signal strength can be measured in terms of microvolts (uV). Out on the airways, FAA considers a good, usable signal to be in the order of 25 uV or greater, but the typical nav receiver has a flag sensitivity setting much lower—perhaps 1.5 to 3.0 uV. Thus, the receiver will "unflag" with a signal that is much weaker than FAA considers usable. If the signal were accurately processed at this weaker level, there would be no harm done, and the receiver could be considered "better" than necessary.

But what guarantees that the signal is processed accurately? Unfortunately, not much. Avionics technicians we consulted said they check the accuracy of a VOR receiver at a much higher signal strength—typically 100 uV. At this much higher setting, the receiver is carefully aligned to produce an accurate needle. What about the signal strength area between unflagging (3 uV) and the bench-check level (100 uV)? What is the accuracy of the receiver in this area? "We have no spec for that," said the chief technician at a large avionics shop.

A spokesman for a well-respected avionics maker told us that the standards for radio design demand that the needle be accurate within three degrees. But he conceded that because bench procedures do not call for it to be checked, there is no actual guarantee.

The FAA considers 25 uV on the airway to be a good signal, but as we've discussed above, there is reason to believe this doesn't all get to the radio. In fact, it is likely that less than half gets to the radio, perhaps not even that much. Thus, it is possible to be in an area where the VOR delivers 25 uV to the airplane, but the receiver only "hears" a signal about 12 uV—assuming a good antenna system in good repair. Let the antenna system be of poor quality, poorly installed or poorly maintained, and the signal at the radio could be only 6 uV. This is enough to unflag the indicator, but may not be enough to safely use for navigation.

Clearly, the airplane whose antenna system delivers the highest proportion of the received signal to the radio will be the safest to fly in this situation. At Port Angeles, flight checks up and down the centerline of the airway yielded signal strengths consistently above 25 uV; but in the area about 55 nm east of Tatoosh and four miles south of Victor 4, signal strength dropped to 20 uV and below.

What to Do?

The Northwest Region staff had discovered the problem, but it was larger than they could handle. They raised the MEA on Victor 4, and it was now considered safe, but their airways standards gave them no reason to raise the MEA, and the standards still allowed another Victor 4 to be created. "We believe that other regions may be unknowingly experiencing the same conditions," they said in their letter to Washington.

Likewise, general aviation aircraft essentially the same as the Warrior continue to be delivered, and tens of thousands continue to fly the airways. There is no standard for antenna systems that insures safe navigation if pilots should happen to encounter another Victor 4.

A 402B Trims a Tree

When the avionics go to lunch, pilots should not always expect a flag to warn them of the unserviceability of a unit. Often, the nav portion of the radios slips quietly away without saying good-bye. Sometimes, this is compounded by erratic indications which, when displayed at the proper time, can mimic correct operation and lead a pilot down some path other than the glide path.

Take, for instance, the case of the 10,000-hour pilot of a Cessna 402B who ended his ILS approach to Newport News, Virginia, by trimming the top off a pine tree after problems with his ILS receiver put him off-course for the runway. Although the aircraft was substantially damaged after hitting the tree, the pilot was able to maneuver for another approach.

The accident occurred after an uneventful night flight from Louisville, Kentucky, where the pilot had departed and proceeded IFR to Newport News. The en route phase of the flight proceeded with no problems.

Upon arrival in the vicinity of Patrick Henry International, the flight was vectored onto the ILS for Runway 7. The pilot reported he was established on the localizer and proceeded inbound with the instruction to report JEFCO intersection, the final approach fix for the ILS. Visibility at the time was three miles in fog and rain.

As the flight progressed down the ILS, troubles began. The pilot, temporarily distracted while searching for the landing light switch, looked up to find that his instruments told him he was far to the right of the localizer. Believing that the aircraft had wandered while he was distracted, he began correcting back to the left.

The tower controller, following the flight's progress down the approach by radar, informed the pilot that he had missed the report at the final approach fix, that he was on a one-mile final and cleared to land. Twenty-five seconds later, the tower controller saw the Cessna deviating from the localizer and told him "you're way off course, pull up." At about the same time, the pilot broke out of the clouds and found the runway was off to his right side, instead of to his left as the localizer needle indicated.

The Cessna was vectored to a point about eight miles out, where the controllers tried to align him with the runway once more. Meanwhile, the weather was deteriorating with the ceiling dropping to zero, the sky was obscured, visibility was one and a half miles in rain and fog.

The pilot again reported he was established on the localizer, and the approach controller handed him back to the tower. At this point, the Cessna began to wander. The on-board ILS equipment was feeding him faulty information, leading him north of the localizer. As he descended (the glide slope portion was still good), he followed the indications until he was about a half-mile left of the centerline.

Breaking out of the clouds, he spotted what appeared to be the runway in about the position it should be according to the instruments. Unfortunately, what he saw was a well-lighted highway, the first portion of which had lights along one side, mimicking approach lights. The last portion of the road was lined with lights on both sides. On top of all this, the road was on the same heading as the runway.

The pilot continued his descent, lining up on the road. As he dropped ever lower, he realized his impending mistake and aborted the landing. He was too late, however, and clipped a large pine tree with the right wing. He would later describe the sensation as being like a landing.

Amazingly, the aircraft remained controllable; the pilot climbed out and was vectored around for another approach. Now realizing he had a problem with the nav receiver, he informed the controller, who responded by providing vectors in a manner similar to an ASR approach. This procedure was successful, and the Cessna 402B landed without further incident.

The NTSB investigated this as an accident. Because of the pilot's comments about the unreliable localizer indications, the investigation centered on the ILS transmitter for Runway 7. Despite exhaustive tests, no problems were found with the ground-based units.

Next, the investigator turned his attention to the on-board equipment of the 402. Once again, despite exhaustive tests, no problems were found with the receivers.

The NTSB "probable cause" was the pilot's improper IFR procedure, indicating that perhaps he should have decided that his navs were unreliable earlier in the approach. There is evidence which should have led him to this conclusion, particularly when he broke out on his first approach to find the runway to his right, when the instruments indicated it should be to his left.

George Is a Great Guy, But...

The "follow-through" is just as important in instrument flying as it is in golf or baseball or any other human endeavor in which continued action is required to achieve the desired result. That may be stretching a bit for a comparison, but consider the number of accidents that have resulted from a pilot failing to continue with a good plan at a crucial point in a flight—have you never pulled back the power, dumped the flaps and put an airplane on the runway when you spotted it at the very last minute in an approach? Have you never wished that you had executed a missed approach, as you had planned, and come back for another try?

The decision to go around at the missed-approach point is usually a very sound one, mostly because it's made in the calm, rational atmosphere that prevails before the approach is begun. But when a last-minute attempt to land is seriously complicated by a system failure—or in this case a pilot-induced system problem—the door is often opened for disaster to strike.

Rapidly changing weather had caused a Montana air-taxi pilot to exercise his decision-making skills throughout the flight. While taxiing for takeoff at Billings, he was advised that Miles City—the intended destination—was below IFR minimums, so he changed to a VFR flight plan that would permit him to overfly Miles City on his way to an alternate airport. There was some indication that the weather might improve along the way.

This was a 2,300-hour pilot with considerable time in a Rockwell 680, but only 32 hours of experience in the 690— similar airplanes, but just different enough to provide some distraction at a critical time.

In the vicinity of Miles City, Flight Service advised the Commander pilot that the visibility was three-quarters of a mile—just short of the one-mile visibility the pilot needed before he could begin the approach. But just as the plane was about to overfly the airport, the FSS advised that visibility was now one mile, with a ceiling of 400 feet, sky obscured. The pilot elected to attempt the VOR-DME approach to Runway 22. After flying the initial approach procedures and the final approach course down to the MDA, the pilot engaged the altitude-hold and approach modes of the autopilot and remarked to a passenger that the rest of the approach was just a matter of "flying out the time" to the missed approach point.

At the appointed time, he was unable to see the runway and was about to start the missed approach, when the front seat passenger announced he had the runway in sight. Simultaneously, the pilot spotted the runway threshold almost directly below him. With 5,000 feet of runway ahead, the pilot felt the Commander was in a position for a safe landing, and he pulled back on the power to descend. However, the nose pitched up and before the pilot could react, the plane struck the ground in what he called "a controlled crash."

Investigators said that although the autopilot's status at the time of impact could not be determined, the circumstances and the pilot's statement were consistent with a failure to disengage the altitude hold mode before retarding power. In other words, the autopilot did exactly what it was programmed to do—maintain altitude no matter what, and that meant a violent pitch-up when the power was suddenly reduced.

An autopilot system is a wonderful tool for single-pilot IFR work; but its safe use depends on a complete knowledge of the system and the ability to integrate all of the autopilot controls and functions into the other work that must be done during the final phases of an approach.

Of course, the other wonderful tool is a pilot's firm resolve to continue with the plan he formulated earlier—i.e., a missed approach—when the first try doesn't work out as expected.

Other Aircraft Systems

An instrument pilot who is properly trained, current and proficient in IFR techniques and procedures, and who is also "at home" in the aircraft he's flying should be able to handle any system malfunction that shows up—inflight fires and structural failures excluded. This assumption would include pilot response to a vacuum system failure (reversion to partial-panel techniques), or a total electrical failure (application of the lost-communication rules, even if it means flying a long way to get to VFR conditions), or a "false alarm"—an *apparent* malfunction that proves non-existent after a bit of trouble-shooting.

Having taken fires and broken wings out of the picture, there aren't many systems problems that will cause an airplane to fall out of the sky. But there are occasions when the distraction and confusion that result from a malfunction cause the pilot to momentarily forget aviation's Rule Number One—*Fly the Airplane!* When airplane control is lost, little else matters; and when it happens in instrument conditions, the outcome is rather predictable.

Here are some examples of circumstances in which failures or malfunctions of various aircraft systems played a major role but, in every one, the day would have been saved if the pilot had continued to *Fly the Airplane!*

Unreal Problems

IFR currency and a reasonable amount of instrument experience weren't enough to prevent the fatal crash of a Piper Comanche 250 on an ILS approach to Dekalb-Peachtree Airport, near Atlanta. The pilot had bought the used Comanche several days before the accident. He had a demonstration ride in which he handled the controls but did not log time, then had a one-hour checkout ride in the aircraft, before flying it from St. Augustine, Florida, to his home base at Peachtree.

Three days later, the pilot went on a short flight to eastern Tennessee, logging one hour of actual instrument time along the way. Returning to Atlanta in VFR conditions, the pilot reported a temporary electrical problem; the air-

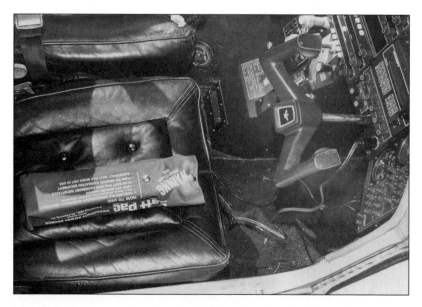

For serious instrument flying in a single, a back-up battery pack is a wise and worthwhile investment.

plane's generator appeared to have failed, and the battery was soon depleted. The radios were inoperative, and he circled the airport while he tried to get the landing gear down. After a short time, the generator problem seemed to solve itself, the gear extended and he landed without incident. The Comanche was inspected by mechanics, who cleaned the voltage regulator contacts and put the airplane on jacks for a gear check, which showed normal operation.

The plane was not flown again for 10 days, when the pilot set out alone on an IFR flight from Peachtree to Anderson, South Carolina. He was 40 miles into the flight when he notified Atlanta Center that he had an electrical problem and wished to return to Peachtree. He was transferred to Atlanta Approach and given vectors for the airport.

The pilot at one point said he had had "a little radio problem" and that he was in heavy rain. After several missed communications due to poor radio quality, he said, "We're just getting a drain on our battery. We'd like as direct to the airport as possible."

The Comanche was vectored onto the ILS Runway 20L approach at about a mile outside the outer marker, but did not acknowledge the controller's 30-degree heading change to get on the approach course. Consequently, he overshot the localizer to the right and the controller issued a correction to the left. The controller then stated, "You can intercept the localizer or give it another try," to which the pilot responded, "Yeah, we'd like to try to shoot it." This was construed by controllers as meaning he would continue the approach.

A ground witness near the outer marker saw the plane at about 100 feet AGL in a shallow left bank with the engine running steadily. A few seconds later, the plane crashed in a steep left-wing-down, slightly inverted attitude.

Investigators found no internal damage in the airplane's generator, and the destruction of the wreckage made it impossible to assess the operation of the voltage regulator. The propeller showed evidence of substantial power at impact. The vacuum pump showed no evidence of failure, and the gyro horizon indicated an attitude approximately the same as the airplane at impact.

The Peachtree weather, given to the pilot before departure, showed a measured overcast at 300 feet, visibility three miles in fog and haze. At about the time of his takeoff, controllers made a special observation of a measured 200-foot overcast, 1-1/2 miles in fog and haze. Two minutes prior to the crash, conditions had deteriorated to an indefinite ceiling of 100 feet, sky obscured, one mile in fog, barely above ILS 20L minimums (DH 200 feet and 3/4 mile visibility).

His logbooks show that the pilot had 742 total hours, eight of them in type, mainly the flights mentioned above. He had logged 47 hours of simulated IFR and 76 hours of actual weather experience, and had flown two actual instrument hours in the past 90 days.

The pilot had enough total experience to recognize an electrical malfunction but, for unknown reasons, he was unwilling to either continue toward better weather (his destination was VFR throughout the episode) or unable to overcome the distractions and concentrate his attention on airplane control during the approach procedure.

Pilot, Know Thyself...and Thy Airplane

Many instrument-rated pilots fly for hundreds of hours without having to execute a missed approach procedure; but when the need arises, it is certainly the wrong time to be in an unfamiliar airplane. A slight difference between airplanes can produce a serious error, as evidenced by a crash that occurred at Duncan, Oklahoma. A 3,000-hour commercial pilot died along with three passengers when a Beech B55 Baron struck the ground three miles west of the Halliburton Airport during an apparent missed approach in blowing snow.

The pilot was cleared for the VOR approach, and was not heard from again; terrain and distance from ATC facilities precluded radar or radio contact during the approach. It was assumed that the pilot was unable to see the runway in time to land, and started a missed approach. The procedure dictates a climbing right turn, but evidence led investigators to believe that the Baron entered a left turn during the attempted missed approach, which may indicate an error or distraction on the part of the pilot. However, this was not their most significant finding.

They discovered that the Baron had impacted wings-level but in a steep nose-down attitude. Furthermore, the landing gear was retracted, the wing flaps were fully extended, and the elevator was trimmed substantially nose-up. There was nothing to indicate any pre-impact malfunction of the airplane, and the evidence thus suggested that a stall had occurred just before impact.

Investigators checked further, and found that this pilot normally flew a Cessna 421. But that airplane was down for repairs, and according to a previous agreement, he borrowed another company's B55 Baron for the trip in question. Over a long period, the pilot had about 40 hours in this model aircraft, 15 in the last three months; it is reasonable to believe that he had never executed an actual IFR missed-approach in the short periods he had flown it.

The Cessna 421 flap system is operated by a lever that can be moved to any position between full up and full down, whereupon the flaps move to the selected position and stop. On the other hand, the Baron flap selector has up, down and neutral positions. If the

_Whether single or twin, dual vacuum or single, partial panel
practice should be a regular part of training._

selector is moved down, the flaps will travel downward to their full
limit unless the pilot stops the system by putting the lever in the
neutral position. If, when cleaning up the airplane during the initial
part of a missed approach, a pilot merely moved the lever from the
down to the neutral position, the flaps would not retract.

The evidence suggests that the pilot inadvertently or by choice
had selected full flaps on the approach, but did not get them
retracted on the missed approach because of this feature of the
system. He may well have been distracted by the resulting poor
climb performance of the Baron, rotated the nose too high, and let
the airspeed bleed off, resulting in a high-power, full-flap stall. And
in most airplanes, that is an abrupt event. He was probably in IMC
until just before impact, a period of time no doubt filled with
confusion, disbelief, and inability to recover.

The Day All the Systems Came Unglued

Most pilots depend on their aviation education and training to
sustain them in times of trouble; but here's the story of one who may
have trusted to a higher power to keep things in order—even

though the systems malfunctions that did him in must have been blatantly obvious.

The pilot—an evangelist—suffered serious injuries when his light twin struck trees and crashed slightly left of and just inside the outer marker on an ILS approach to Panama City, Florida. There was an apparent engine failure just before the 7:25 p.m. crash.

The instrument-rated pilot had reported 3,500 total hours on a recent FAA medical application, with 200 hours in multi-engine airplanes.

The investigation showed that during the day of the accident the pilot had departed Panama City and flown to Meridian, Mississippi, then to Pensacola, Florida, with the entire trip filed and flown under the instrument rules. The pilot was returning alone to Panama City when the accident occurred. Local weather included a 600-foot overcast, visibility five miles.

Curiously, the approaches made during the other flights of the day began as pilot-conducted procedures, but all of them were changed to vectored or Airport Surveillance Radar approaches at the pilot's request. When he began the ILS approach at Panama City, controllers noticed the airplane was drifting off-course, followed immediately by the pilot's request for a radar surveillance approach. The controller refused because of workload, so the pilot requested a radar vector from his present position to the airport, with a letdown to the ILS Decision Height. This also was refused, and the crash occurred shortly thereafter. The injured pilot was not located until 11:40 a.m. the next day; he had been trapped in the wreckage, unable to reach a portable transceiver in the cockpit.

Investigators said that at one of the stops on the day of the crash, the pilot had found he could not start the plane's right engine. A mechanic found that the fuel transfer pump in the right tip tank was not receiving electrical power. Tracing the problem, the mechanic found that inside the right engine cowling, a wire to the fuel pressure sensing switch—which enables the fuel pump—had separated. He soldered the wire, and the engine started without difficulty. Examination of the wreckage after impact showed the same wire separated, even though the cowling was intact.

The pilot told investigators he was on the approach and at the outer marker he activated the fuel pumps; the right engine immediately quit and there was a rapid yaw to the right. He told investigators the "cage" light of the attitude gyro came on, and he believed the gyro may have shown an inverted attitude prior to impact. However, the plane struck the pine trees at a nearly level attitude.

Witnesses reported that the airplane had very few avionics devices in working order, and an ADF and the portable transceiver may have been the primary avionics. In addition, investigators said, a scan of records disclosed that the pilot had been involved in two previous aircraft accidents, once with his family aboard. How many warning flags does a pilot need?

IFR Departures

T he folks who put the words on standard
instrument departure charts have taken to
indicating either "pilot nav" or "vector" so
that aviators know what might be expected when a SID is part of
an IFR clearance. That's a qualified statement—"what might be
expected"—because in most cases, flights are radar vectored out of
the terminal area anyway. But just in case of a communications
problem, a "pilot nav" SID would indicate that there is a published
route to be followed; a pilot on a "vector" SID would proceed direct
to the navigational fix for which the vector was originally intended.

Although SIDs are primarily convenience items for use in high-
traffic terminal areas, they also provide obstruction clearance—
and there are few things more important to a departing IFR pilot.
The architects of the airspace system have not forgotten the less
busy airports, however. Once again, instrument departures fall
into either of two categories; those with published instrument
departure procedures, and those without. (Of course, when there's
an air traffic control facility on the field, departure instructions will
be issued just before takeoff; for the rest of this discussion, we'll
assume that there's no local ATC involved.)

In the case of published departures, pilots are given specific
headings or courses, minimum altitudes and sometimes minimum
climb rates to guarantee obstacle clearance on the way to an en
route airway segment with its own minimum altitudes.

The question of what it takes to safely leave an airport that does
not have a published instrument departure procedure became

paramount when a Cherokee pilot attempted to leave William T. Piper Memorial Airport in Lock Haven, Pennsylvania, and join an airway to the east. Certainly not "mountain-bound" by western standards, Lock Haven is nonetheless located in a valley which requires either careful navigation or remarkable climb performance—or both—if a pilot is to safely clear the terrain. This pilot had neither going for him; he crashed into a ridge about five miles east of the airport.

The pilot told investigators that as he was preparing to leave, he was handed the phone by another pilot who had concluded a briefing with Williamsport FSS. Rather than receiving assistance, the Cherokee pilot was told "curtly" to call back, he said. But when he did, he could not get through again to the Williamsport FSS. Consequently, the pilot called the Wilkes-Barre FSS, obtained a briefing and filed an IFR flight plan.

As the pilot related to investigators, "I had filed Victor 232, Milton VOR, Hazleton VOR, direct Hazleton [his destination airport]. The clearance I received was direct Swiss intersection, V232 Milton, direct Hazleton. Accepting the logic that Wilkes-Barre has more expertise in issuing clearances for this area and that this routing would keep me clear of the terrain, I accepted it without question. The assigned altitude was 4,000 feet MSL."

The pilot took off from Runway 9, climbed to 1,000 feet AGL and entered the base of the clouds. At 1,600 feet MSL, he radioed Wilkes-Barre Approach for initial contact. He said communications were poor and he intended to make another attempt at 1,700 feet. However, just as the plane reached that altitude, he heard a loud bang and then the Cherokee was crashing through trees and falling down the far side of a ridge.

Seriously injured but a survivor, the pilot told investigators it was only after the accident he learned that "the local rule of thumb for an IFR departure is to turn to a heading of 070. This isn't known to transients, nor is it posted anywhere. With an MEA of 4,000 feet for V232, a climbout on runway heading didn't seem like an impossible maneuver. Either a prescribed IFR departure should be posted, or a recommendation of no IFR departures from the airport should be posted."

According to the Airman's Information Manual, airports may fall into three categories for departure under IFR. An airport may be covered by a Standard Instrument Departure procedure (SID), it may be governed by a published IFR departure procedure, or it may have no published procedure. The manual states that "each pilot, prior to departing an airport on an IFR flight, should consider the type of terrain and other obstructions on or in the vicinity of the airport and take the following action:

• Determine whether a departure procedure and/or SID is available for obstruction avoidance.

• Determine if obstruction avoidance can be maintained visually or that the departure procedure should be followed.

• At airports where approach procedures have not been published, hence no published departure procedure, determine what action will be necessary and take such action that will assure a safe departure."

Do What the Chart Says...But Do It Carefully

The airport at Bishop, California, is located in a valley at 4,120 MSL with mountains up to 13,000 feet in the general vicinity. Bishop VOR (BIH) is located on the field. It's no surprise that Bishop has a published instrument departure procedure, which requires pilots to "Climb visually within 2 nm of Bishop Airport to cross BIH VOR at or above 8,000 feet, climb SE-bound on R-140 to 10,000 feet, climbing left turn, proceed to BIH VOR crossing at or above 12,000 feet, continue climb..." and goes on to describe on-course procedures. The initial instruction to climb visually within two miles of the airport was a central issue in the lawsuit generated by a Convair 340 crash several years ago. The accident also brought attention to an alleged "black hole" phenomenon affecting a takeoff, rather than an approach, as is usually the case.

The flight had arrived late at Bishop from Burbank, California, and contacted the Tonapah, Nevada, FSS to file an IFR flight plan for the return trip. At 8:21 p.m., the crew contacted the FSS to advise the flight was taxiing for departure on Runway 12 and requested its IFR clearance. At 8:24 p.m., the flight radioed, "Climbing VFR over Bishop, awaiting clearance."

The Convair crashed four minutes later at the 6,100-foot level of a 6,300-foot ridge, 5.2 nm southeast of the airport, hitting in a right-wing-low attitude. It was clear that the crew did not stay within the two-mile circle, but why?

Evidence showed that the night was moonless and high clouds obscured even starlight, and although the airport was lighted, there were no lights in the mountains anywhere in the area to the east. This has all the ingredients of a "black hole" situation, wherein pilots making night approaches have extreme difficulty in judging attitude and height above ground when the runway is the only lighted reference, surrounded by dark water or terrain.

Calculations showed that the Convair could have accomplished the two-mile spiral, but it would have been a steep climbing turn near the limits of the plane's performance.

Plaintiffs argued that the VFR circling departure is simply unsafe and that FAA was negligent in preparing it. They developed evidence to show that the FAA official who wrote the procedure never even visited Bishop Airport, never looked at an aerial photograph of it, nor made any inquiry of local people about conditions at the field. Rather, the official drew up the procedure in an office at Oakland using a 1949 Coast and Geodetic Survey chart and an outdated FAA manual. Moreover, the visual departure procedure was put into service without FAA pilots making a night departure from Bishop, and the airline was given FAA approval to operate Convairs in and out of Bishop day and night without any night proving flights to show that it could be done safely in that size airplane.

The FAA defended itself on all points. According to lawyers representing the agency, its officials often write procedures without ever visiting an airport; but they do so according to rigorous standards that insure the procedures are safe. The FAA does not have the resources to flight-check each and every procedure under all conceivable conditions, nor is it necessary if a procedure is safe under all existing guidelines.

"The FAA can't possibly check every procedure for all conditions of visibility, lighting, cloud cover and so on," said the government lawyer who plead the FAA's case. "And if they were to make a night check, what night should they pick? Would a moonlit night be sufficient, or a dark night? If it's a dark night, would they have to wait for cloud cover or could they just use a clear night?" She also said use of a 1949 topographic map has no bearing on the matter, since "the rocks don't change very much over the years." Moreover, the procedure was circulated among local lightplane pilots using Bishop Airport before it was put into effect, and the FAA received no objections, she said.

The government attorney was convinced that the crew of the Convair never intended to use the procedure. She said they were tired, behind schedule and probably only completed a right climbing turn from the runway and intended to continue southward down the valley before topping the mountains and going on course.

In perhaps the oddest twist in the case, the FAA argued that the pilots were not conducting part of a published IFR procedure because the procedure says "climb visually" and because the pilots radioed that they were "climbing VFR over Bishop." In essence, the FAA contended that the pilots were not following the procedure, or not flying it correctly.

The plaintiffs contended that the crew was trying to follow the procedure but could not because of the disorienting effects of the "black hole" situation, that the procedure is unsafe in such conditions, and that the FAA was therefore negligent in preparing and authorizing use of it.

The NTSB was unable to come up with a probable cause for this accident, and although the circumstances are somewhat unique— sharply rising terrain, super-black night, the requirement for a precise turn in a very limited space—it underscores for all IFR pilots the importance of positional awareness, especially when a mistake spells certain disaster.

When You Can't See Where You're Going

There's an apparent paradox at work when you consider the takeoff minimums that keep air carrier pilots grounded, while pilots operating under Part 91 can depart legally in zero-zero conditions. This seems a super-long leash for the non-professionals, but it's just another manifestation of the general FAA safety policy; pilots who are not selling tickets can do just about whatever they want as long

A takeoff into low-IMC is the most demanding part of any instrument flight.

as they exercise a reasonable standard of care, but when the operation becomes commercial in nature, the tighter regulations impose a much higher standard of care for the protection of the flying public—after all, that's the FAA's charter.

The instrument takeoff is still a legal procedure and still is practiced by some pilots. Whether it is a prudent and safe procedure is another question—we believe it simply is not. One can sympathize with a pilot wanting to get through the first couple of hundred feet of ground fog to VFR conditions above, but nothing is quite as safe as waiting for the fog to burn off.

Two problems are obvious. First, having lifted off from the fogged airport, suppose the pilot experiences engine failure? He now must return to an airport he can no longer see, or crash as best he can.

Second, the instrument takeoff demands reliance on a single instrument—the directional gyro. In virtually every other IFR task, an instrument pilot cross-checks this primary instrument with some other indication. But in an instrument takeoff, the wet compass is the only secondary heading reference available, and it is far too inaccurate to be useful.

Everything rests with the DG. But it doesn't tell the pilot if the airplane is drifting left or right; a five-degree error—just one increment on the DG scale—would put a wheel off the edge of the runway, assuming a 700-foot ground roll and a 150-foot wide runway. A series of smaller deviations, if summed in the same direction, could do the same thing—and all this assumes the pilot lines up perfectly at the start, and the DG works as advertised.

A flight instructor piloting a Piper Cherokee 235 at Sierra Sky Park in Fresno, California, found he could trust all but that last assumption when he crashed during an attempted instrument takeoff. If a pilot can have everything going for him, this pilot seemed to. He got a preflight weather briefing, and on his way to the airport confirmed visually that conditions were VFR above a fog layer. Visibility on the field was 20 feet.

The sod runway was 400 feet wide, providing lots of leeway for heading inaccuracies. He carefully positioned himself, set his DG to 300 degrees (the runway heading), and commenced the takeoff roll.

He kept the DG pegged on 300 degrees as he rolled down the runway. His description: "At that point the aircraft ran off the runway to the right with the right wing dropping suddenly, leading me to think the right landing gear had been sheared off by collision with the runway markers. I could not believe this was possible, as I had not deviated from a 300-degree heading. The aircraft then bounced over an embankment and the left wing collided with a steel pipe at the crest of the embankment. The plane came to a stop beyond the pipe after turning around 90 degrees and sliding sideways." The pilot emerged with minor injuries. His airplane had described a path that was roughly 30 degrees off course.

What had happened? It was too simple to be believed—just another application of Murphy's Law. When he set the gyro, the pilot didn't notice that the black plastic cover over the instrument panel was catching the DG knob, preventing it from uncaging.

The airplane came to rest just short of a residential area—the injuries to innocent people might have been catastrophic. On top of that, the left wing had been sheared off by a telephone pole, which

Before taxiing onto the runway, give the gyros plenty of time to spin up and check them carefully with taxi turns.

would surely have increased the pilot's injuries had the airplane hit it dead center. These are items to ponder when contemplating a "legal" instrument takeoff.

Stopping in Time

A simulated instrument takeoff (using a hood and with a competent instructor in the right seat) is a great training exercise; it really teaches an IFR student to concentrate on heading control. But given all the things that can go wrong during an instrument takeoff in actual zero-zero conditions, there is usually a better alternative.

Two Pennsylvania pilots managed to escape injury when their Aztec veered off the runway during an aborted IFR takeoff; weather conditions at the time were sky obscured, ceiling zero, visibility less than one-eighth of a mile, temperature 31, dewpoint 30.

Pilot in command was an ATP in the right front seat, accompanying a commercial pilot who was flying from the left seat. They were about to depart on an IFR flight plan flight when the accident occurred.

In response to questions about possible pitot system icing, the PIC told investigators that the plane had been hangared the night before, a preflight check confirmed that the pitot heat was working, and that the pitot heat was on during the takeoff roll. Despite these precautions, the right side airspeed indicator never showed a reading, and when the PIC realized that visual contact with the runway could not be maintained, he chose to abort the takeoff. During the deceleration, the Aztec left the runway, with little damage and no injuries.

A reason for the apparent pitot system failure could not be found; the left-seat pilot told investigators her airspeed indicator had reached 45 miles per hour.

Whoops, we almost forgot; the runway was covered with four inches of snow. *That* will do a number on takeoff acceleration!

Air Traffic Control Procedures

When the ATC system is good, it's very, very good. But when it's bad, it's horrible. Most pilots perceive that the ATC system works best when it works its own way, and that anything out of the ordinary (unusual routings, bad weather, equipment failures) raises the chances of things going wrong.

That conclusion was largely confirmed by a study of reports made to NASA's Aviation Safety Reporting System. When all the toys are in place, when nobody has bizarre routing requests, and when the weather is not a factor, things are hunky dory. However, when operations begin to get abnormal, controllers show a rise in errors, particularly in the areas of monitoring and communications.

To understand the nature of the problem, researchers reviewed a total of 135 incidents recorded in safety reports from Air Route Traffic Control Centers. The researchers identified certain "primary contingency factors" which resulted in the controller's subsequent error. These primary contingency factors included:

- Weather deviations
- Special use airspace
- Weather
- Routings
- Communication functions
- Radar data processing
- Narrowband/broadband operation
- Broadband operation

- Sector boundaries/status
- Aircraft system functions
- Radar/radio coverage

"It was observed," the researchers noted, "that many of these factors (53 percent) relate to situations occurring near the boundaries between air traffic control facilities." Feeling they were onto something, the investigators pursued the question of what was going on—or failing to go on—at these boundaries. In 39 of the 71 incidents the handoff or pointout was late or omitted, and there were two other incidents in which several controllers in a row failed to note an aircraft was operating with some discrepancy from its posted flight plan.

Turning to controller performance at other than sector boundaries, the researchers divided controller performance deficiencies into three categories; planning (figuring out where planes should be as they transited the sector), monitoring, and information transfer. There were 64 incidents occurring entirely within a sector, under one controller's jurisdiction. Of these 64 foul-ups, 42 percent were judged a result of planning deficiencies. Another 35 percent were monitoring errors, and 12 percent were information transfer problems (11 percent could not be classified).

Under contingency conditions, the controllers made fewer planning errors but more monitoring and communication mistakes than when they weren't responding to a contingency. The reason suggested is that "the planning function is seen as a primary task by controllers whereas monitoring and communications are secondary." This, in turn, is linked to the typical controller's personality profile. "Controllers as a group tend toward strong, dominant, aggressive, self-confident personalities. Planning is an active function, monitoring a passive one." The researchers also note that "The planning task under such conditions may be so compelling as to interfere with the accomplishment of other duties perceived as secondary."

These findings have implications for the current overhaul of the National Airspace System. In determining the type and quantity of automated equipment which controllers will have at their disposal for moving planes through the sky in coming years, planners are taking a close look at what tasks are the most error-conducive and attempting to automate those. This would leave controllers more time to do what they do best, and reduce the risks of fatal mistakes for both controller and pilot.

Positional Awareness Again

A flight instructor and his student were killed when their plane struck the side of a canyon northeast of Burbank, California, after controllers placed it on a vector toward high mountains. The Skyhawk was apparently in IFR conditions at 3,000 feet when the crash occurred; mountains in the area rise above 6,000 feet.

NTSB investigators said they are looking at a number of factors, including controller communications, the workings of the ATC radar system, and the flight instructor's lack of flying experience in the Los Angeles area.

The accident came on the instructor's first day of flying after being transferred from Florida to join his flight school's operation at Santa Monica, California. He reportedly had spent a week at Santa Monica a month earlier, and had been briefed on the area.

The CFI and his student had taken off from Santa Monica earlier in the day and conducted a training session without incident. But when they arrived back at Santa Monica, they were unable to land because a Bonanza had landed gear-up. They were delayed more than an hour and a half waiting for Santa Monica to re-open, and during the delay the weather closed in, prompting the CFI to file an IFR flight plan. The route of flight would take the Skyhawk near Burbank Airport, and controllers issued radar vectors due to other traffic departing Burbank.

The C-172 was given a series of vectors by the Burbank Approach controller. After the last heading of 080, there were no further communications from or to the aircraft for about eight minutes.

At about the time of impact, the Los Angeles Approach "stadium sector" controller queried the Burbank Glendale sector controller about a target he observed about eight miles east of Burbank. The Glendale sector controller replied that he did not know anything about the target and would check with the valley sector controller. The target disappeared shortly thereafter.

Controllers Make Mistakes, Too

There are a number of situations during an IFR flight when pilot and controller responsibilities overlap, but as the Airman's Information Manual (and common sense) makes perfectly clear, the pilot must always be the one with the responsibility and the authority to undo what someone else may have done. This shows up on occasion when a controller is distracted and "loses his place" when directing a stream of traffic in IFR conditions. One such incident turned into an accident at Jacksonville, Florida and provides a lesson in knowing where you are vertically as well as longitudinally.

The pilot and his passenger were killed when their Piper Cheyenne hit the ground short of the runway while on an ILS approach to Jacksonville after an hour-long flight from Hampton, Georgia.

The flight was routine until the approach commenced. The pilot had been radar-vectored by Jacksonville Approach toward the final approach course, then was given instructions to fly an intercept heading, join the localizer and maintain 3,000 feet.

The approach procedure calls for crossing the outer marker at 1,900 feet, and as the Cheyenne neared the OM, the controller issued a clearance for the ILS but, due to an oversight, failed to instruct a descent from 3,000 feet to the appropriate altitude.

The flight was then handed off to Jacksonville Tower; the local controller advised the pilot that he was about half a mile outside the marker and to descend as published on the approach chart. The pilot acknowledged.

Communications transcripts show the controller handling the flight became concerned when he noticed it hadn't descended from 3,000 feet prior to the OM. As the plane's descent began, the controller told his supervisor he wanted to issue instructions to terminate the approach: "I'm going to break him off. It don't look good to me." The supervisor replied, "What's he (the pilot) say?" When there was no word from the pilot, the approach was allowed to continue.

Still concerned, the controller noticed that the plane was gaining on an Eastern Airlines Boeing 727 that was ahead of it on the approach. "He's 40 knots faster than Eastern," said the controller. "Just wait and see what happens," replied the supervisor. "See if it works out."

It worked out, all right. The Cheyenne struck a tree-covered swamp about a mile short of the runway and 347 feet right of the centerline, with the landing gear down and the flaps retracted. Initial impact was in a roughly level attitude at a high rate of speed. Weather at the time included a measured 300-foot ceiling, visibility one mile in fog and light rain.

Investigators threw out the possibility of an encounter with the wake turbulence of the B-727, and plots of the radar data from the Cheyenne's transponder showed that the pilot had apparently tried to salvage the approach from a position well above the glide path. In that situation, it is necessary to descend much faster than normal, which can easily lead to a deterioration of airspeed control, and as the increased groundspeed requires an increased rate of descent, the problem feeds on itself. The controller's mistake notwithstanding, the pilot in command remains responsible for noticing the problem and doing whatever is necessary to correct it. In this case, a missed-approach was clearly called for.

The Pilot's Emergency Authority

A recent Administrative Law Judge's decision dismissing an FAA Order to suspend an airman's certificate offers some interesting insight as to the extent to which a pilot may rely upon emergency authority when deviating from ATC clearances.

The pilot had recently purchased a new Piper Malibu with all the bells and whistles, but no right-side flight instruments. On the day of the incident, the departure airport was IFR; the pilot was cleared to depart Runway 10, turn right to 290 degrees, climb to and maintain 2,000 feet.

He later testified that shortly after lift-off he entered instrument conditions, retracted the gear and throttled back to climb power. The Departure controller acknowledged radar contact and repeated the clearance to turn right to a heading of 290 and maintain 2,000 feet. Shortly thereafter, the pilot noticed that his attitude indicator showed a steep right descending turn, while his turn-and-bank and HSI indicated a wings-level climb.

In attempting to resolve this disparate attitude information, the pilot delayed his right turn and inadvertently climbed through his assigned altitude to 2,800 feet. When queried by the controller as to his heading and altitude, the pilot responded in a voice several octaves above his normal frequency that he was "all screwed up," followed shortly by "I'm in trouble." The controller ignored the indirect plea for help and berated the pilot for not returning to his assigned altitude. During this period the Mode C readout showed a maximum of 2,800 feet.

Ironically, once the pilot returned to 2,000 feet he was immediately cleared to 3,000 feet; by this time, he had resolved the attitude dilemma and was relying solely on the turn-and-slip, altimeter, VSI and HSI.

The Malibu was then turned over to a different departure controller who cleared him to his requested cruising altitude. During this further climb he broke into the clear at 4,000 feet. Within minutes the attitude indicator realigned itself and the remainder of flight to Olathe, Kansas, was uneventful, VFR all the way.

The pilot had the attitude indicator and vacuum system inspected in Olathe, but other than some minor contaminants found in the vacuum lines, no other discrepancies were found that would have affected the proper operation of the attitude indicator.

Shortly after this incident the pilot received a letter of investigation from his local FSDO, to which he responded without hesitation. He stated that the failed attitude indicator had so preoccupied his attention that he inadvertently climbed through his assigned altitude, exercising his emergency authority pursuant to FAR 91. Next the pilot received a formal letter of investigation for an enforcement action, and was invited to discuss the matter with a representative of the FAA's Regional Counsel. Having been led to believe that this was just one of those friendly fireside chats, the pilot was surprised to be confronted with a meeting that could be more accurately described as an inquisition with both the FSDO representative and the attorney furiously taking notes on everything the pilot had to say in his own defense.

But the FAA was dissatisfied with the pilot's version of the inci-

A momentary glitch in the Malibu's AI caused the pilot to climb through his altitude. The FAA pursued a violation but the NTSB dismissed the case, arguing that the pilot had properly exercised his emergency authority.

dent; a formal Order of Suspension for a period of 30 days was issued for "careless operation" and "deviating from an ATC clearance."

At the formal hearing before an Administrative Law Judge, the FAA rested its case after introducing the transcript of communications and a tape recording of the communications in which the pilot acknowledged that he had violated the altitude clearance. Under the rules for such hearings, and the precedent by which Administrative Law Judges determine who has the burden of proof, the burden of defense then shifted to the airman to prove that he in fact had a legitimate emergency and that the emergency was not of his own making.

The most persuasive evidence was the pilot's moving testimony as to the seriousness of losing his primary attitude indicator moments after liftoff and in close proximity to the ground. This testimony, of course, was corroborated by the excited communications on the tape recording. In further support of his position, the pilot introduced the definition of an emergency found in the *Airman's Information Manual*—"any situation of distress or urgency." The pilot also asked the judge to take judicial notice that the FAA, by its own regulations pertaining to the requirements for obtaining an instrument rating, lists as a simulated emergency any instrument malfunction. It was also argued that it was all the more an emergency since the flight test guide for an instrument rating does not even require recovery from unusual attitudes following the loss of an artificial horizon. The pilot also referred to Webster's definition of emergency as being "a sudden unexpected happening, specifically a perplexing contingency or complication of circumstances or a sudden or unexpected occasion for action, exigency or pressing necessity."

The FAA argued that since the pilot never declared an emergency, no emergency in fact existed and that by inference the loss of the attitude indicator was a fabrication used in an attempt to explain away the altitude violation. This put the credibility of the witness at issue, a matter for the Administrative Law Judge to examine and rule upon.

In furtherance of the pilot's defense, Section 9 of the ATC Controller's Manual was introduced; this section requires that a controller who may be in doubt as to a particular situation treat it as an emergency and provide the pilot with all available assistance.

The ATC Supervisor testified that when he listened to the tape, he could not detect any indications of distress or urgency and that the mere use of the words "I'm all screwed up" and "I'm in trouble" did not indicate any problems. He further testified that in his view, the pilot's declaration that he was "in trouble" meant that the pilot simply was acknowledging that he was in trouble with the FAA for violating the altitude restrictions.

Witnesses for the FAA acknowledged that although there was a technical violation of the altitude clearance, no other aircraft were affected with regard to separation standards; controllers in adjacent sectors were immediately alerted to the situation and resolved potential traffic conflicts.

After deliberating on the evidence, the Administrative Law Judge found that the pilot's testimony was credible and that indeed an emergency condition existed which justified the pilot's deviation from his altitude clearance, even though he didn't formally declare an emergency by the use of the words "emergency," or "mayday." The judge dismissed the Administrator's Order of Suspension.

You might think that the ATC system would respond to a situation like this by providing the pilot with affirmative and immediate assistance. But it doesn't always happen that way, and the lesson to be learned from this particular case is that if there is any doubt in your mind, resolve that doubt by declaring an emergency to assure that the applicable requirements of the ATC Manual are implemented by the controller. Without the formal declaration, the matter of emergency-or-not is left to the subjective determination of the controller.

There was a time where you could believe the statement, "I'm from the FAA and I'm here to help you." Those days are history. Discussions with FAA representatives—whether by telephone, in writing or at "informal conferences"—are not for the purpose of providing pilots with counsel and guidance to become safer users of the airspace system; they are for the purpose of gathering evidence for use against the pilot if and when an enforcement action is instituted.

Another thing this particular pilot learned is that GA aircraft are not required to have failure indicators for flight instruments such as are required in transport category aircraft. An attitude indicator failure can be quite insidious, and when it occurs within a few hundred feet of the ground, only a few seconds are available to detect and correct the situation. At least one Malibu now flying has a newly installed set of copilot flight instruments and a dual vacuum system.

ATC Radar—Capabilities and Limitations

When it comes to detecting aircraft, there are many reasons why an airplane might not show up on ATC radar screens. One problem is related to the radar cross section of the plane and causes some aircraft, such as a jet airliner with lots of reflecting metal surfaces, to show up very well on radar, while a small aircraft produces a much weaker return.

Temperature also affects radar reflectivity; the hotter the object (like the turbine section and exhaust of a jet engine), the better the reflection. A light aircraft, like a Cessna 172, is comparatively cold. The heat produced by a jet engine helps compensate for another factor—the aspect of the aircraft. If the aircraft is pointing directly toward or away from the radar antenna, there's less area to reflect the radar pulses. For jets, the hot portions of the engines provide excellent reflectors. But small aircraft don't show up as well in this situation; the amount of energy reflected may drop below the radar's detection level.

ATC's ASRS-4 radar uses huge antennas with transmitter outputs of up to five million watts. Such radars have a range of more than 150 miles.

Another factor that affects all targets, regardless of size, is tangential speed. At a certain range, targets moving at just the right speed and course appear to have no velocity, and the radar's computer program deletes such zero-velocity targets as clutter. Thus, they do not show up on the screen.

With all these factors working against small general aviation aircraft, it's a real crapshoot as to whether they will show up at all. A Cessna 172, pointing the "right" way, moving in the "right" direction at the "right" range may as well be a Stealth fighter—it will not show up on current ATC radars.

Finally, an aircraft can be rendered "radar-invisible" by atmospheric temperature inversions, which also contribute to "beam bending." The radar beam enters the inversion and can be bent down toward the ground (shortening the radar's range and filling the screen with clutter), or the beam bends up, skipping over the area it's supposed to cover. This phenomenon is also related to the time of year; cold weather tends to bend the beam down, warm weather bends it up.

The Transponder Factor

All of these problems can be overcome with radar beacons, or transponders as they are more commonly known.

The beacon system operates in much the same way as raw radar, but a second antenna is mounted above the main radar antenna, and as they rotate, the beacon antenna sends out transponder query pulses. The replies from airborne transponders are processed to provide distance and azimuth, plus additional coded information such as the transponder code, altitude and groundspeed.

In its most comprehensive configuration, ATC controllers should be able to "see" all transponder-equipped aircraft, therefore making traffic separation much easier and more positive. This is part of the reasoning behind many of the airspace changes the FAA is promoting. (By the way, have you ever wondered from whence cometh the term "squawk?" It goes back to WWII days, when transponders were known to the military as IFF—Identification, Friend or Foe—and the system was so secret it required a code word. "Parrot" was chosen because that bird replies with whatever it hears; parrots squawk, and the term found a permanent place in the aviation lexicon.)

But even transponders experience some technical problems that limit their effectiveness, one of which is "synchronous garble." Under certain conditions—for example, when two aircraft are flying on converging or near-parallel tracks—their transponder signals may show up as one aircraft or several, because the electronic signals from each transponder overlap the other.

This problem grows in proportion to other factors. In high-density areas, more aircraft translate into a greater chance for overlapping transponder replies. The number of querying radars also adds to the problem. In an area like New York, an aircraft may be continuously tracked by dozens of radars, each one querying, each one reading each reply; the chances for overlapping signals increase remarkably.

Although processing and software changes have overcome some of the synchronous garble problem, it has not been licked. "I don't care if they admit it or not, it's still a problem," says one radar expert, who also believes this will be a problem for the proposed

TCAS systems. With TCAS units querying transponders and generating replies, the incidence of synchronous garble will increase both for the radars and for the airborne indicators. While the radars can cope with the problem to a certain degree, it remains to be seen whether future TCAS units will be able to overcome this potentially fatal flaw.

Mayday! Mayday! Now What?

When an engine quits over rugged terrain, or when ice starts accumulating on the airframe, or when the pressurization system suddenly can't hold its breath at 25,000 feet, a pilot needs all the help the ATC system can provide. Unfortunately, there's evidence that in many instances pilots will have trouble getting that help because of communications problems.

Lulled by the security of two radios, even pilots who are not in contact with ATC assume they can just pick up the mike and yell for help with a reasonable expectation of having their request heard, understood and acted upon. But that ain't necessarily so.

A study of reports to NASA's Aviation Safety Reporting System yielded 168 cases of emergencies, and in 31 percent of those cases, the pilots had communications problems during their time of travail. Surprisingly, the problem was rarely a question of being out of radio and/or radar range; only three cases cited limitations in that area, with another three describing equipment failure.

Leading the parade of communications failures was "lack of coordination" by ATC. This is the situation where Controller B somehow doesn't get the news from Controller A about what the situation is, or where two controllers work at cross-purposes. These failures resulted in two delays in landing, one extension of an emergency, two landings without proper emergency equipment standing by, five traffic conflicts and one operation without an appropriate clearance.

Next on the list was controller inattention; this is a hazard not so much to the emergency aircraft as to the others under ATC control at the time. The emergencies studied produced a total of 14 instances of traffic conflicts, 10 of which did not involve the aircraft experiencing the emergency.

Of the 52 cases where communications problems were identified, 24 pilots experienced an exacerbation of their problem as a result of poor communications, and 28 had the dubious honor of

Implemented in early 1990, TCAS has suffered some initial glitches. The manufacturers and the FAA say this is to be expected and that software fixes should correct minor problems.

finding themselves with an entirely new and different hazard on their hands because communications broke down.

Sharing the Blame

While it sounds like ATC is always at fault, that's far from being the case. There were five instances in which the pilots were inexplicably reluctant to declare an emergency, perhaps out of fear of getting enmeshed in the bureaucracy upon landing. Of course, what they gave up was an enhanced chance of making a landing in one piece, because the declaration of an emergency changes the rules for ATC controllers and allows them to put more resources to work.

Is there such a thing as a good emergency? In a sense, there is— at least from a communications point of view. The study found an unexpectedly low frequency of communications failures in emergency situations where VFR-only pilots got themselves into IFR conditions. "It may be inferred," said the researchers, "that controllers, for whatever reason, are particularly sensitive and adept with regard to these emergencies."

The other good news from the study—"There is no evidence in these data that pilots, even with minimum experience levels, are not familiar with proper emergency radio procedures."

More About ATC Radar

Since the inception of the radar-based air traffic control system, pilots have come to rely on the ATC system to steer them clear of traffic and keep them out of mid-air collisions. When the weather gets rough, pilots expect ATC to help them avoid the worst of it.

Many pilots have also discovered the system has limitations. Other aircraft don't always show up on the radar scope, with or without a transponder. Weather depiction on FAA radars is poor at best, nonexistent at worst. And relying on ATC to help avoid collisions or severe weather has led pilots down the garden path to disaster.

New radars are in the works for ATC, and will offer some striking improvements over the systems currently in use. Unfortunately, the FAA will not be using some of these features and may abuse others in a perhaps misguided attempt to "improve" the safety of terminal control areas. And pilots will not be getting radar services which are noticeably different from what they got before.

Great Expectations

One of the services pilots fully expect is to be separated from other traffic. Indeed, the FAA proclaims that this is their primary mission for ATC—keeping airplanes apart in the sky. This directive over-rides all other activities for controllers.

Yet, as illustrated by several disastrous mid-air collisions and hundreds of near misses, ATC is not always able to provide this service. Airplanes continue to come together under all kinds of conditions. Although the FAA might argue that the big problem is non-participating aircraft, the accident records (and near-miss reports) show this isn't always so.

Pilots also expect to be provided with weather information from ATC radar, and vectors around hazardous weather when it shows up. Again, the accident records show this doesn't always happen. The explanations lie in equipment limitations and the human limitations of the controllers, as well as the FAA's philosophy of air traffic control.

Got That Traffic?

A world full of transponders, all squawking and encoding, should give controllers a handle on everyone's position and altitude. But, as with everything else, there are limits.

Perhaps the biggest limitation is the controller himself. After all, dozens of little blips and their data tags on an 18-inch radar scope can be more than a little overwhelming. Targets and data tags start crowding each other, making it hard for the controller to tell which tag goes with which blip. A mid-air collision between a Skywest Metroliner and a non-transponder-equipped Mooney M20 over Kearns, Utah, illustrated just such an occurrence. The NTSB considered the problem of overlapping data tags as a contributing cause to this accident.

Nevertheless, controllers have options for dealing with this. Some of the problems can be handled by their equipment, and aircraft that are squawking and encoding but are not in that particular controller's airspace can be filtered off the screen. Thus, aircraft which are above or below the controller's airspace need not be displayed on his scope.

Another option is to suppress all the VFR transponder codes (i.e. all transponders squawking 1200). The controller's handbook states that "During periods when ring-around or excessive VFR target presentations derogate the separation of IFR traffic, the monitoring of VFR code 1200 or code 1277 may be temporarily discontinued." This option, when exercised, can be a real kicker from the VFR pilot's point of view. He can never tell whether his transponder is doing him any good or not. He may be flying along with his transponder set on 1200, squawking and encoding. But if the controller is suppressing the VFR codes, that pilot may as well have his transponder turned off.

Recently, one of our readers shared his experience with a situation like this. Although squawking 1200 and encoding his altitude, he was almost hit by a Boeing 737 that came from behind and below. Obviously, the airliner was showing up on the controller's scope, but our reader's Beech 23 Musketeer evidently was not.

Even though the number of targets on his scope can be reduced, controller overload can still be a major problem, one that can spill over into adjacent sectors. For example, a near miss involving a Beech 99 commuter airliner and a Piper Navajo. The Beech 99 was en route from Boston to Burlington, Vermont, at 8,000 feet, the Navajo was flying from Bangor, Maine, to Teterboro, New Jersey, also at 8,000 feet. Both aircraft were on IFR flight plans. While they were in airspace controlled by the Manchester position of the Boston TRACON they nearly collided. From the NTSB report: "The aircraft passed with zero feet lateral separation and approximately 200 feet of vertical separation as the Beech 99 nosed over to avoid an imminent mid-air collision. The radar controller was aware of the pending conflict, yet was busy issuing coordination instructions and holding aircraft in his airspace as a result of adjacent sector saturation. The controller did issue a descent clearance to the Beech 99 when the aircraft were approximately one mile apart."

Weather, or Not

Pilots have come to expect weather avoidance service from air traffic controllers. The expectation is that the controller will be able to see weather in his area on radar and provide vectors around it. This used to be true when the old broad-band radars were used, but now the answer is a heavily qualified "maybe"—and as future radars come on the scene, the answer may well turn out to be "no."

The radar equipment currently used at FAA Air Route Traffic Control Centers performs poorly with regard to weather depiction. The reasons for this lie in two characteristics of Center radar—the inability to tilt the antenna (and find storm height) and the type of signal the radar sends out. The result is that weather returns on the center controller's scope are depicted as a series of lines. These lines point back at the radar site which is illuminating the storm, and provide little information about the weather.

Center radar can only provide three levels of weather return— none, some, and heavy. "None" does not mean there's nothing there, it only means the radar doesn't see anything there. "Some" means the radar sees something which might be rain. "Heavy" means the radar sees something which is probably rain.

None of these is an accurate indication of what kind of weather is present; it could be a light sprinkle, it could be moderate rain, it could be a tornado-producing thunderstorm, or it could be a temperature inversion with no precipitation at all.

The point is that *pilots can't rely on controllers to detect heavy weather and provide vectors around it.* A pilot flying a Beech Duchess found out just how important this limitation can be.

He had taken off from Tulsa, Oklahoma, bound for Little Rock, Arkansas, one afternoon when thunderstorms were forecast along the route of flight. Although the Duchess was equipped with weather radar, the unit was not working for this flight.

After taking off from Tulsa, the pilot contacted departure control. "They put us on a heading of 120 degrees with a climb to 6,000 feet, which was aimed at the southern end of a large black thunderstorm with visible lightning every 4 or 5 seconds," he later reported. The pilot tried several times to tell the controller of his impending entry into the thunderstorm, but got no answer.

"Finally, I told approach I was turning south to avoid the weather. They said they wanted me to turn to a heading of 050 degrees. I declined, as that was the direction of movement of the thunderstorms, and it was black as far as the eye could see in that direction."

The controllers told him he would have to fly some 50 or 80 miles south to skirt the weather. He accepted this and was handed off to Fort Worth Center, whereupon he was cleared direct to Okumulgee, then direct to Fort Smith, then to pick up his flight-plan route. When the pilot advised the Fort Worth controller he did not have weather radar and would like vectors around the storms, the controller handed him off to Memphis Center.

He repeated his request for weather vectors to the Memphis Center controller, who declined to help, saying their radar "was not weather capable." The pilot was told to proceed direct to Fort Smith.

"We were in the clouds in light to moderate turbulence and light rain," the pilot later reported. "I was steering visually towards the lighter areas and maintaining the airway. About ten miles east of Okumulgee, we got into about four minutes of heavy rain and I slowed to 130 knots. In the middle of this four minutes, there was about 90 seconds of hail. The noise was like being in a pillbox on Omaha Beach on "D" day. After the encounter with the hail, we could see paint damage and some dimpling of the outer wing panels and leading edges of the T-tail. After landing, we discovered we had $15,000 worth of damage."

TRACON: Yesterday Lives On

Center radars operate in a mode known as "narrow-band," which eliminates many of the problems experienced with the older "broad-band" radar, but Terminal Radar Approach Control (TRACON) facilities are still operating with the old "broad-band" systems.

While aircraft on the radar scope in the ARTCC show up as

slashes for transponder targets, and a cross or pinpoint for non-transponder targets, the TRACON radar scope shows these differently. All aircraft, whether transponder or not, show up as rather fuzzy bands parallel to the range lines on the radar scope. Those aircraft with transponders (or manually identified by the controller) get a data tag and a letter of the alphabet imposed on the target—all others remain fuzzy bands.

But TRACON radar sees and displays weather better than the ARTCC radar. The old broad-band system, using analog signals, detects weather (and other atmospheric phenomena) very well. The terminal controller has a much better picture of its shape and size than the controller at an en route center, and he can cut vectors much closer than his ARTCC counterpart.

There are some problems here as well. When the weather starts getting intense, it can effectively mask radar returns from aircraft. Most importantly, the controller has almost no control over whether he sees the weather on the screen or not. Unlike the center controller who can select weather at will on his screen, the TRACON controller is at the mercy of the system engineer who physically turns the weather on and off on the controllers' scopes. When weather returns are starting to clutter the scopes for most controllers, he can reset the equipment to show either the outline of the weather or eliminate it entirely from the screen.

Thus, when it comes to weather information from a TRACON facility, pilots should expect even less than they get from center controllers. When the weather starts getting intense (just the time pilots need those vectors around the storms), it's very likely the system engineer will take the weather off the controllers' scopes.

New Radars Coming

In response to these and other problems with the current radars in use at ARTCCs and TRACONs (like maintainability—vacuum tubes are getting harder to come by all the time), the FAA is shopping for new radars. The proposed radars have some fabulous features—unfortunately, controllers and pilots may never get to use some of them.

ARSR-4 radar will be installed at ARTCCs around the country, quite literally, because plans call for the new radars only along the coasts and the borders. The FAA feels it has enough ARSR-3 radars (the current model in use) in the rest of the country.

The new radars will be serving the FAA for air traffic control and the Air Force for air defense; the dual role includes some features which could be very useful to the FAA.

For example, the Air Force wants the radar capable of determining the height of an aircraft within 3,000 feet at a distance of 175 miles. Additionally, the radar must be able to detect a target with a radar cross section of one-tenth a square meter—considerably smaller than a Cessna 150. With this kind of capability, controllers should be able to detect aircraft with or without a transponder, determine the approximate altitude, and issue traffic advisories as needed. Unfortunately, the FAA controllers will not be using this capability, because the altitude measurement is not accurate enough when compared to an encoding transponder.

At present, ATC radars have no height-finding ability, leaving controllers with not even approximate altitudes on non-transponder aircraft. The new radars, although not as accurate as transponders, would at least provide some information which the controllers could use to help keep IFR traffic away from the non-transponder targets.

But the FAA refuses to consider using this ability; instead, it is considering the purchase of another piece of radar equipment which was developed for the U.S. Army. The radar has height-finding capability, a range of about 50 miles, and the FAA plans to cover a 40- to 60-degree slice of the Los Angeles TCA with it during a test program. The objective is to enhance collision avoidance and detect TCA violators. If the new radar is successful, it may be installed at other TCAs around the country. It will not replace Mode C, but it will be useful in tracking TCA violators.

The Shame of NEXRAD

The new radars for the ARTCCs may also be able to detect weather. General Electric is one of the firms vying for the contract, and the equipment it proposes can detect weather, look through it to see targets inside the storm, and provide turbulence information like the most advanced weather radars. According to GE spokesmen, this information could be supplied to the controller's display with some software changes.

But the FAA will not be using any of this information; the FAA is waiting for NEXRAD, a weather radar which the agency believes will provide the weather information the controllers need.

NEXRAD (Next Generation Radar), is being designed for the kind of weather detection the FAA believes controllers and pilots need. It will have Doppler signal processing to detect windshear, micro-bursts, and turbulence. It will display precipitation in varying levels, showing the intensity of a given storm much like airborne weather radars do.

NEXRAD may live up to its name, but not in the way the FAA intended. At the rate the schedule is slipping, it may indeed be generations from now before NEXRAD is actually in the field. Research on NEXRAD began in 1979 as a joint project between the FAA and the National Weather Service (NWS). James King, former chairman of the NTSB, testified before Congress in 1981 that the technology for NEXRAD has been around "since the 1960s and could have been adopted considerably earlier."

Initially, it was planned that NEXRAD would start being deployed in the field during 1985. By 1983, the FAA was telling Congress that NEXRAD was delayed, and planned deployment was pushed back to sometime between 1988 and 1992. Recent interviews with FAA personnel indicate the schedule has slipped again by a year or more, putting NEXRAD back to the mid-1990s. Part of the reason for the slippage has been infighting between the NWS and the FAA over such things as the colors of the displays; the FAA has one set of colors it wants to use, and the NWS has a well established color-coding system which it prefers.

Meanwhile, pilots are stuck with little if any help in avoiding severe weather. While NEXRAD is in development, the FAA refuses to examine other alternatives. Controllers will remain incapable of vectoring pilots around the most severe weather until NEXRAD gets out of the lab and into the field.

ASR-9: TRACON Grows Up

Improved radars are also in the works for TRACONs. The Airport Surveillance Radar 9 (ASR-9) will be a real improvement; it uses a digital radar signal that has several advantages over the current analog system. With the current ASR-7 and ASR-8 radars, an analog signal is broadcast and reflected back from aircraft, then it is processed into a digital signal so the transponder information can be gleaned from the returns. This digital signal is then reprocessed back into an analog signal so it can be projected onto the radar scopes. A computer matches the transponder information with the

NEXRAD, although behind schedule and over budget, may significantly improve ATC's ability to detect weather and issue aircraft radar vectors to avoid it.

radar scope presentation. On the whole, this is a very cumbersome system.

The new ASR-9 will eliminate all this processing. The digital signal also allows use of Doppler techniques to eliminate clutter and weather returns from the controllers' scopes. Tangential targets will also be detectable, enabling the ASR-9 to track a target as it enters the tangential speed range, follow it through that range and pick it up again on the other side—current radars can't do this.

There's a down side however; the ASR-9 will not be able to deal with beam bending and temperature inversion. Although project engineers have been trying to overcome these problems through signal processing, one engineer said "We don't know if this will work 100 percent."

A separate weather channel incorporated into the ASR-9 will offer contoured weather radar presentation in two levels (moderate and heavy) on a separate scope, but it will not detect or depict turbulence or windshear.

Future Shock

Although these new radars will have capabilities beyond those of present systems, ATC radar services will not be enhanced. In fact, the FAA indicates that it may eliminate radar altogether, opting for a totally transponder-based system.

A spokesman at the Atlantic City Technical Center told us there have been studies conducted by the FAA on eliminating radar from the ARTCCs. If this does come to pass, the implications are grim. Aircraft without transponders (or with their transponders turned off) will not be detected in any way. This opens the door to an increase in the number of near misses and mid-air collisions, because controllers will be unable to provide traffic advisories concerning these "non-participating" aircraft—the controllers won't even know they are there.

Under this kind of a system, not having a transponder won't mean just reduced radar services. It will mean no radar services. It will also mean big troubles for aircraft whose transponders fail in flight. With current equipment, a failed transponder causes the aircraft to show up as a primary skin-paint return, but the aircraft would disappear from the scope under the new system.

In some ways this can be viewed as the culmination of the FAA's push to get all aircraft to carry transponders. In response to legislation adopted by Congress several years ago, the FAA issued a rule that significantly reduced the airspace available to non-transponder aircraft. Another rule requires Mode C transponders for operations in Airport Radar Service Areas (ARSAs). With these two rules in place, some areas will be totally off limits to non-transponder equipped aircraft. Since the coverage area of some ARSAs extends from the ground to 10,000 feet, and transponders are required above 10,000 feet, non-equipped aircraft will be forced to go around the ARSA.

Mode C and Avoid

No matter what equipment requirements the FAA puts into effect or what radars the FAA buys, pilots will still be ultimately responsible for avoiding collisions and severe weather. While the new radars for the ARTCCs might be able to detect a beer can 20 miles away, this is no guarantee that a Cessna 172 at the same range will be seen. Even having a transponder is no guarantee of protection—and being on an IFR flight plan doesn't reduce the risk of a mid-air collision to zero.

The new radars will have some fabulous features, but their impact on radar services for pilots will be minimal. Many of the same limitations which plague the current systems will continue to be problems with the newer equipment—and those limitations have been fatal to some pilots in the past.

The best protection against mid-air collisions and severe weather will continue to be the human eye. Looking out the window, a human pilot can detect all manner of hazards that the best radars in the world can't detect.

Human Factors, Failures in Air Traffic Control

The LearJet pilot couldn't have been any too happy after a 25-minute ground hold, the purpose of which was never fully explained to him. That was the first of several mistakes the controller made, and the pilot added a couple of his own. Unfortunately, it was the pilot who paid the price.

The instrument clearance from the Palm Springs, California, airport called for a right turn after departure, direct to the Palm Springs VOR (PSP), then via the PSP 051 radial to Twentynine Palms VOR. In practice, pilots were rarely called upon to make the greater-than-90-degree turn directly to the nearby VOR; normal procedure was a series of vectors onto the PSP 051 radial.

After departure from Runway 30 the pilot continued straight ahead, and the controller didn't notice that anything was wrong. For reasons all too human, several opportunities to avert disaster were missed.

For example, three minutes after takeoff the pilot and the controller exchanged this information:

N12MK: Okay, are we cleared to 051 to Twentynine Palms?

PSP: That's affirmative, sir. I had a change of route there....Maintain niner thousand. I'll keep you advised.

Perhaps the pilot, flying in IMC, had some inkling he was headed for trouble. Ninety seconds later he voiced his concern:

N12MK: Where is Mike Kilo going after this? We're maintaining nine on a heading of 310. What's our clearance from here?

PSP: Yes, sir. You can expect further clearance crossing the 20 DME fix, sir.

The controller later told investigators he thought the pilot said "...a heading of 030," but on the phone to Los Angeles Center he said "It was like he might have said heading maybe 230, see, I'm not sure, but he was cleared on the 051 radial." The controller failed to realize, long after the fact, that a heading of 030 would never intercept the 051 radial after a departure from Palm Springs.

By now, Center controllers began to get a data block on their radar scopes. In ninety seconds all of the information was on the screen and Center knew the Lear was northwest, not northeast, of the airport, level at nine thousand, traveling at almost 300 knots and heading for mountainous terrain. Center wasn't talking to the Lear, and by the time the interphone operator reached the Palm Springs Departure controller, the Lear had aluminum-plated a mountainside 24 miles from the airport.

Palm Springs, California, was at that time a non-radar terminal, which meant that controllers had to keep track of aircraft and separate them by using the information on flight strips—an archaic procedure at best.

This may also have been an unfamiliar situation for a pilot accustomed to operating out of large airports where radar control is the rule; he could easily have assumed his departure was being observed electronically, particularly since on the inbound leg he'd had radar coverage from Center until late in the approach when terrain blotted out electronic detection. It was a deadly assumption that compounded the pilot's initial failure to make the right turn after departure.

What really went wrong, however, was a very familiar tendency on the part of all participants in the ATC system to assume things are as they should be. This unbelievably strong bias in favor of normalcy caused the controller to ignore information that the flight under his control wasn't where it should be. And of course the pilot failed to realize where he was.

The pilot said he was on a heading of 310, almost 90 degrees to the 051 radial, and an impossible intercept course. The controller didn't notice.

The pilot later reported reaching a nonstandard fix. The controller didn't notice.

Too much time passed before the pilot reported crossing the VOR. The controller didn't notice.

While increased radar coverage can help, it is not the solution. As long as pilots and controllers start out with the assumption that everything's okay, errors like this will continue to occur. The only antidote for this human factors problem is to train ourselves to assume everything *isn't* okay.

IFR Approach Procedures

Surely, no analyst of aviation safety would be surprised to discover that the approach phase is the most hazardous of all instrument operations. And with good reason, because that's the time when the objective of the flight is to get as close to the ground as possible in weather conditions that prohibit visual contact until the very last minute—and sometimes not at all.

With "home base" very close at hand, and probably driven by a strong urge to get out of the clouds and into visual conditions, instrument pilots often succumb to the temptation to go just a bit below the minimum altitude, to fly just a few seconds longer in hopes of seeing the runway, or circle to land in very low visibility. The accident records are filled with examples of these and a wide variety of other attempts to beat the system.

Barring negligence on the part of IFR procedure designers or massive errors in navigation signals, there's one thing that instrument pilots can count on; if IFR procedures are flown as published, there's no way you can hit anything. "As published" means strict adherence to the altitudes, courses, speeds and distances that appear on the charts, or instructions which are supplied by radio. The IFR regulations cover virtually every circumstance you might encounter in the airspace system. There's always an out.

No Shortage of Illustrations

Because *Aviation Safety* has reported on the problems of instrument approach procedures in great profusion over the years, and because the material covers such a wide scope of IFR operations, we will present them without categorization. Our intent in this section of *Command Decisions* is to provide you with a series of memorable experiences of other pilots, with the hope that whenever you see a similar set of circumstances setting up on an IFR flight, you'll remember what happened to someone else and take steps to get yourself out of harm's way.

Confusing Lights?

An Aero Commander 690A, a twin turboprop, had been dispatched from Toledo, Ohio, to pick up cargo at Cadillac, Michigan. Both occupants were killed when the airplane struck trees and then the ground during an instrument approach to Cadillac's Wexford County Airport. The pilot in command was an ATP, accompanied by a private pilot who was along for the ride, and who was not considered a copilot.

The Aero Commander was cleared for an NDB approach to the airport, and as far as controllers were able to tell, executed the approach without incident—the plane was heard to fly over the airport on the initial pass, then was observed approaching the airport from the south. Weather at the time included a 600-foot overcast, visibility one mile in rain. The impact site was south of the airport, with the Aero Commander on a northerly heading. Investigation revealed evidence of ample power on the engines at impact, and there was no sign of any malfunction of the airplane.

Investigators noted that microphone-activated runway lights had recently been commissioned at the airport, but it was not clear whether the pilot of the Aero Commander had received a NOTAM to this effect. The flight reportedly was the pilot's first night approach to Cadillac. Significantly, witnesses said the airport lights never activated.

It was also noted that a Christmas tree farm was located just south of the crash site and a series of bright lights had been set up there for harvesting the tree crop at night.

It appears that this was a matter of mistaken identity, and of course the approach procedure guarantees don't extend to obstacle clearance when attempting to land on a tree farm. This accident also underscores the vital importance of studying the facilities (including runway layout, lighting systems, etc.) before attempting an approach—especially a totally unfamiliar one—and executing a missed approach when things don't look the way they should.

Problems with Texas Fog

A commercial pilot with a considerable amount of experience in his Bonanza came to grief during an NDB approach to Pampa, Texas. Current and qualified for instrument flight, the pilot had logged 6,700 total hours, including 494 in the Bonanza. He had conducted the same approach earlier in the day of the accident, but this time it was in the dark and in weather conditions that appeared to be well below the published minimum.

Fog had rolled into the area shortly before the approach. The Pampa Airport operator had shut down for the night because of the weather, and no wonder; the Amarillo observation (closest to the accident site) reported the sky was obscured, indefinite ceiling zero, visibility one-quarter mile. As if that weren't enough, the remote altimeter setting required an increase in MDA to 600 feet AGL—not much hope of seeing the runway lights through the fog.

The Bonanza was cleared for an NDB Runway 17 approach, and was observed to complete the outbound leg and procedure turn and begin the final approach. When it was not heard from, a ramp check and search was initiated. Investigators said weather was so foggy that the airport manager missed the turn to his airport when arriving for the ramp check.

The wreckage was found several hours later about a mile north of the runway and 150 yards to the right of centerline. It had struck the ground in an approximately level attitude, gear down, flaps up, then bounced back into the air, traveled another 60 feet and impacted steeply nose-down. The nose dug into the ground and the momentum tore off the entire engine and firewall, leaving rest of the airplane to travel an additional 80 feet before coming to rest.

The NTSB investigator said no pre-impact malfunctions were found. Friends of the Bonanza pilot stated that he had been known to fly below minimums on IFR approaches; an instructor had recently counseled the pilot on the subject, telling him there was "just no future in it."

Professionals Have Troubles, Too

All eight occupants were killed when a scheduled commuter flight struck wooded terrain short of Maine's Auburn-Lewiston Airport during an ILS approach. Controllers had observed the flight deviating from the localizer course prior to the outer marker and had issued vectors to correct it.

The flight had departed Boston with a scheduled stop at Auburn-Lewiston before going on to Augusta, Maine. Investigators said the flight went without incident until the Beech 99 arrived in the Auburn area, where weather was recorded as an indefinite ceiling 300 feet, sky obscured, visibility one mile in fog and drizzle.

The initial approach involves overflying the Kennebunk VOR and intercepting the ILS 4 localizer at a shallow angle (the VOR outbound course is 036 degrees, and the LOC inbound course is 040). The controller handling the flight noted that the airplane deviated to the right of course as much as a mile, and it appeared that it would miss the outer marker completely. The response was "negative," so the controller issued a vector that would allow the flight to intercept the localizer at the outer marker.

The pilot complied with the vector, but then overshot the localizer to the left of course, then corrected to the right again as the target left the radar scope, a routine occurrence because of intervening terrain.

The aircraft impacted trees in a wings-level attitude at 320 feet MSL, 4,000 feet short of the runway and 440 feet to the right of the extended centerline, on a heading roughly parallel to the localizer course. Field elevation is 288 feet.

The landing gear was down and the flaps were retracted. There were indications of propeller rotation at impact, and investigators could find no evidence of malfunction of the engines or airframe prior to impact.

The ILS components were flight-checked and found to be working normally. Auburn-Lewiston has pilot-activated approach lights, and a witness said the airport lights were on at the time of the crash. During the approach, the flight had contacted the weather observer on the field and obtained the current weather, without commenting on any problems they might have had.

Cheating Gets Expensive

There are numerous variations in the game of "cheating" on an IFR approach; for example, intentional descents below minimum altitudes. We are reminded of the time an air taxi pilot held on for the ultimate approach—an ILS all the way to the runway in zero-zero conditions. He was still in his 500 fpm descent when the runway rose up and bit him, and he might have gotten away with it, except for a small problem; he couldn't taxi clear of the runway because the landing gear had been driven up through the wings.

Busting minimums can even assume some structure: after a widely publicized commuter crash, 14 former company pilots told the Safety Board that they would be chewed out if they did not bust minimums coming back to home base, and that company management had offered to pay the fine if the FAA caught a pilot doing it. This might have gone on for years if a crew of two pilots (with the company's chief pilot in command) had not hit trees short of the runway one foggy night.

Another favorite variation involves shooting an approach to one airport and breaking out under a ceiling that is perhaps 500 to 1,000 feet AGL, then proceeding "VFR" to the real destination airport—trusting that no clouds have sunk down to the ground and that the pilot knows every hill or radio tower along the way.

And there's the "homemade approach," where local pilots concoct their own approach procedure. Several factors make this a very dangerous stratagem. First, it surely involves an element of hubris on the pilot's part, since it says he knows better than all the FAA professionals how to develop the procedures, plot the navaids and set the minimums. Second, it generally entails using the least accurate navaids (an NDB or VOR) in an area where a truly safe approach might cry out for a full ILS. Third, using the homemade approach will require some surreptitious behavior. The pilot must tell the controller he is doing one thing, while doing another. If something doesn't go quite right, he will have an added burden of embarrassment when he calls the controller and admits he isn't where he should be at all.

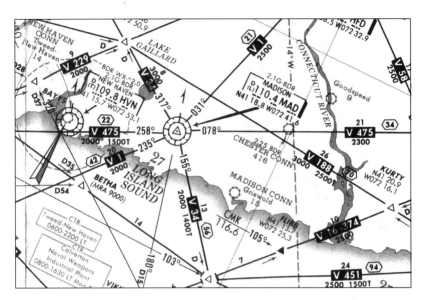

Chester, Connecticut, has a VOR approach, but Goodspeed, a small strip on the Connecticut River, has no approach at all. The Bonanza pilot evidently planned to break out over Chester, then circle to land at Goodspeed.

The NTSB called attention to such a case back in 1975, when a corporate Rockwell 690A crashed near Nemacolin, Pennsylvania. Corporate pilots provided the Board with a copy of an unapproved procedure that was marked "for VFR only," but that had all the elements of a makeshift VOR-DME approach. Nemacolin is at 2,000 feet MSL in a valley where mountains rise to 3,000 feet and higher, and the minimum en route altitude is 5,000 feet. The Commander was cleared for the MEA when the crew reported "some ground contact here" and began to descend.

Considering reported weather, it is doubtful that the flight ever got into VFR conditions. The pilot's last words to the controller were, "If we lose radio contact with you and we make the app...the landing okay, I've got an 800 number to call to cancel it." The twin struck a ridge at the 2,800-foot level, five miles short of the airport, killing all four people aboard.

Although the Safety Board did its best to bring notoriety to the Nemacolin accident, it apparently has not deterred latter-day attempts at imitation. Particularly as more and more light aircraft acquire area navigation equipment, the ability to simply insert a waypoint and instantly put an "outer marker" where none exists has become a godlike power that some pilots just can't resist. But an RNAV waypoint is no more accurate than the VOR from which it is derived, and there is still the question of altitude versus obstructions.

The crash of a Beech V35A Bonanza near Chester, Connecticut, is a case in point. The pilot held a commercial license with an instrument rating and 1,700 total hours, including 223 actual IFR. The airplane was well equipped; it had dual navcoms, ADF, autopilot, flight director, radio altimeter and RNAV.

At 6 p.m., the pilot and his son departed Bridgeport, Connecticut, on an IFR flight plan for Chester, half an hour away. At Bridgeport, the sky was partially obscured, 5,000 overcast, half a mile in light rain and fog, wind calm, temperature 51, dewpoint 50—conditions ripe for the creation of a classic New England fog. Destination weather was equally forbidding: a 200-foot ceiling, sky obscured, one-quarter mile in rain and fog.

To the controller, the flight must have looked straightforward. The plane got to the Madison VOR and was cleared for the VOR approach to Chester. The pilot acknowledged and was not heard from again.

As the pilot later told the Safety Board, his radio altimeter "failed" during his initial descent and that's why he turned it off. At the final approach fix, he descended to 1,200 feet and maintained that until 7 DME, where he decided to abandon the approach and divert to Hartford. "At that moment the aircraft struck trees and subsequently the ground on a ridge I was later told was 485 feet above sea level," he told NTSB in his report. "My conclusion is that because of heavy rain, there may have been a plug of water in the pitot-static system, which gave a false altimeter reading."

Investigators found little to substantiate this story, and much to dispute it. There was no evidence that the altimeter malfunctioned, but it was set to 29.92, which was incorrect—the controller had issued the Bridgeport altimeter as 29.82. Further, the altimeter read 860 feet, which happens to be exactly the MDA for the Chester approach *with the local altimeter setting*—but the pilot never contacted Unicom for that information. A note on the approach plate makes it clear that when the Bridgeport altimeter setting is used, the MDA increases to 1,000 feet.

There's More

The plane crashed a mile and a half north of the airport, whereas the missed approach procedure dictates a climbing right turn, which would lead one south of the field. And though Chester is on the Madison 076 radial, the crash site was on the 066 radial, with the flight director course set on 065. It seemed apparent that the pilot had not conducted a proper approach—or missed approach—to Chester. But why?

There were still more coincidences. Neither of the nav radios was tuned to Madison, but one was found tuned to Hartford VOR, and an RNAV waypoint using a 175-degree radial and 12-mile distance was entered. This waypoint lies on Goodspeed Airport, a small uncontrolled strip along the Connecticut River about five miles northeast of Chester. The plane's 066-degree course, when extended, also led directly to Goodspeed Airport, where the pilot was renting a tiedown. The field is paved and lighted, but it has no approved instrument approach. A possible conclusion is that the pilot never intended to land at Chester at all, but merely used its approach as a ruse to get to Goodspeed.

The pilot said there is no dispute that he intended to go to Goodspeed, but only if he could have broken out VFR at Chester. He said he was conducting the approach to Chester, but did not intend to go below 1,200 feet—and that's what the altimeter said when he crashed. He said he would have conducted the proper Chester missed approach, but never got the chance.

As for some of the odd radio and instrument settings, the pilot said these could have been changed either at impact (his son's head struck the panel), or later by unauthorized people. On this score, he noted that the day following the crash he discovered personal items missing from the plane and had to post a guard to be sure the avionics would not be stolen.

The pilot sincerely believed his barometric altimeter malfunctioned. He said just prior to the flight in question he had flown from Hartford-Brainard, Connecticut, and shot an ILS into Bridgeport.

During a long, smooth final approach, he said, the altimeter stuck for a time at about 1,100 feet, then suddenly jumped down to 700 feet. Despite this, he picked up his son and took off for Chester, since he also had a radio altimeter. But during the descent for the Chester approach, he said that instrument began "pinging" intermittent altitude alerts. Since he had heard this could happen when a signal bounces off heavy precipitation, he turned it off to eliminate the annoyance.

The pilot said the lesson he learned from the accident was that "Never again will I make an approach without a dual redundant system for every critical instrument, such as two independent barometric altimeters." Perhaps readers can imagine other lessons as well.

Another Case of Mistaken Identity

Possible mis-identification of ground lights led to an accident in which a Cessna 177RG undershot an ILS approach to Runway 14 at Republic Airport in Farmingdale, New York. The pilot (a CFI) and his passenger (a private pilot) escaped with only minor injuries.

> The weather was bad along the entire route of flight from Keene, New Hampshire, with ceilings under 500 feet and visibilities less than a mile in rain, drizzle and fog. In addition, there was a convective SIGMET in effect near the route, and flight precautions had been issued for moderate turbulence below 10,000 feet due to strong winds. In his written statement, the passenger stated that at Keene, the "visibility was good, however it was a bit foggy. We were on top at 2,000 feet."
>
> An IFR flight plan had been filed, and the Cessna proceeded south at 6,000 feet along the planned route. After passing the Hartford VOR, the passenger/pilot, who was doing the communicating, acknowledged a clearance direct to the Deer Park VOR for vectors to the ILS at Republic Airport.
>
> The weather at Republic was even worse than had been forecast. The ATIS reported an indefinite ceiling of two hundred feet, sky obscured, visibility one-quarter mile in light rain and fog, with winds from 80 degrees at seven knots. Decision Height for the ILS Runway 14 approach at Republic is 250 feet, and one mile visibility is required.

While being vectored for the approach, the flight requested the weather at Teterboro, New Jersey, and got a report of similar conditions; indefinite ceiling three hundred feet, sky obscured, visibility one-half mile in light rain and fog, with winds from 70 degrees at 10 knots. The pilot chose to attempt the approach at Republic.

The Cardinal pilots were told to contact the tower four miles from the outer marker. Some difficulty was encountered, and it took several tries before communications were established. The tower controller told them to report reaching FRIKK, the locator outer marker, 3.9 miles from the runway threshold. When they did, he said "Report the approach lights in sight, lights are up full bright, cleared to land Runway 14." This was the last communication with the aircraft, and it disappeared from radar less than two minutes later.

In his written statement, the pilot said "I heard [the passenger] report FRIKK inbound when we received the outer marker beacon light. At approximately 350 feet I saw the approach lighting system and told [the passenger] to report the lights in sight. At approximately 300 feet I noticed that the glide slope flag had appeared. I immediately pitched the aircraft up and simultaneously added power to initiate a missed approach. The next sound I heard was the impact of what sounded like branches hitting the aircraft."

The passenger said "While on the ILS approach I called out the altitudes beginning at 500 feet, 400 feet, 300 feet and 200 feet. No attempt was made to arrest the descent. I did not see any lights on the ground. Shortly after that we hit the trees." It is not clear whether the passenger made the radio call "lights in sight," but no such transmission was received by the tower.

The airplane came to rest in Bethpage State Park, at an elevation just under 200 feet and several thousand feet from the runway. The Cardinal was demolished, but there was no fire.

Based on the prevailing visibility at the time and the location of the crash site, it's unlikely that the pilot actually saw the runway approach lights. Exactly what he saw is also unclear. An investigator who visited the site said, "There's nothing out there that looks like approach lights. I don't see how he could have seen them—he hadn't even made it to the middle marker." A last anomaly is contained in the pilot's written report; he stated that he had the approach lights in sight, yet his decision to go around was based on the loss of the glide slope signal.

Quitting While You're Ahead

Some instrument pilots are undaunted when they hear the destination reporting zero-zero. Weather can change, and weather observers can be pessimistic, so why not just shoot the approach anyway? If it doesn't work out, the alternates are still above minimums, and if it does, there's no motel room or rental car to pay for.

There's nothing wrong with this notion if the pilot is current and competent, and if he has the discipline to quit at the MDA unless he gets the threshold environment clearly in sight. If he does not have the discipline, his commencing the approach is like a junk food addict trying to eat just one potato chip.

All these points are chillingly illustrated by the crash of a Cessna 172 at Ithaca, New York. The pilot had been flying regularly for seven years and had 694 total hours, with 103 hours of actual IFR time; he was current and qualified for this flight.

> The pilot got a weather briefing five hours before the flight. The record does not show specific observations, but the general outlook was "IFR with flight precautions for low ceilings and visibilities, and for frequent moderate icing in clouds and precipitation." At 1:39 p.m., the pilot filed IFR flight plans from Bridgeport, Connecticut, to White Plains, New York (about a 20-minute flight), and from White Plains to Ithaca, the plane's home base.
>
> Skyhawk N5161R was still 100 miles away from Ithaca when controllers began to discuss what to do about the flight, in view of the weather. New York Center was handling the aircraft, when Elmira Approach suggested the pilot be told that Ithaca was still "WOXOF"—indefinite ceiling zero, sky obscured, visibility zero in fog—just about as bad as it gets. Three minutes later, the Center controller told the pilot that Ithaca was "below minimums," and requested his intentions.

The pilot replied, "Six One Romeo will probably go right ahead and shoot the approach anyway." The Center controller discussed this with Elmira Approach:

New York: He says he wants to shoot the approach anyway. We told him it was WOXOF. Is he allowed to do that?

Elmira: He wants to shoot it anyway?

NY: Yeah.

ELM: Yeah, we can give it to him. I got no problem with that.

NY: I can't—I don't have a problem with it. I'm just curious if there were, ah, you know....

ELM: What it is—if they're below minimums, if he should land up there they can write him up if they want to.

NY: Oh, okay...That's what he wants to do right now, anyway.

ELM: Yeah, well, he wants to be written up then, I guess.

Shortly thereafter, the pilot asked the controller for permission to leave the frequency to check weather conditions in the area.

NY: Six One Romeo, that's approved, report back. I just checked with Elmira. They said that Ithaca is WOXOF.

Pilot: Say again about Ithaca?

NY: Ithaca is below minimums right now—WOXOF—indefinite ceiling, sky obscured, zero visibility in fog.

Pilot: That doesn't sound very promising. I'm going to check around, see where else has got better weather. Where I am right now is completely VFR.

The Skyhawk was about 50 miles east of Binghamton, or about 80 miles from destination, at 8,000 feet. The record does not show the pilot's conversation with Elmira FSS, but the weather briefer logged it as a request for observations at Syracuse, Rochester, Elmira and Buffalo. The briefer gave reports on all but Syracuse, whose observation had not come in, and suggested to the pilot that Rochester's weather seemed the best. Rochester was reporting partial obscuration, measured ceiling 900 feet overcast, visibility two miles in light snow and fog. Buffalo was measured 300 broken, 1,300 overcast, visibility one mile in light snow and fog. Elmira was indefinite 400 feet, sky obscured, visibility 1-1/2 miles in fog. They were all above minimums. The record does not show an observation being requested for Binghamton, the closest ILS-equipped airport at that point. At 4:50 p.m., the pilot came back on Center frequency:

Pilot: Six One Romeo is still—like to go to shoot the approach at Ithaca. If that doesn't work out, I think I'll try either Binghamton or Elmira.

NY: Roger. Binghamton is below also. Elmira is open right now.

Pilot: I just got Binghamton as 400 overcast, one (mile). Is that, ah, minimums?

NY: It may have just come up. As of ten minutes ago, it was below.

Pilot: Okay. Well, it will take me awhile to get there anyway.

At 4:57 p.m., the pilot reported "Six One Romeo is starting to pick up some ice" and requested a descent to 7,000 feet. The controller told him to stand by, but only 20 seconds later, the pilot said, "Okay, 61R is out of the ice at the moment." They discussed lower altitude assignments, but the pilot said, "Seven thousand feet looks like a good altitude for the time being."

The flight continued without incident until 5:21 p.m., when Center effected a hand-off to Elmira Approach about 12 miles southeast of Ithaca. The pilot was given the current Ithaca weather: "Indefinite ceiling zero, sky obscured, visibility zero in fog, temperature and dewpoint 35, Runway 32 is the active. The wind 310 at four, altimeter 30.01, expect ILS approach." The pilot was cleared down to 3,200 feet on vectors to intercept the localizer for an ILS to Runway 32.

The pilot got position reports at seven miles from the marker, five miles out ("cleared for the approach") and one mile from the marker, when the controller said:

ELM: Cessna 5161R, a mile from VARNA, radar service terminated, contact Ithaca Tower 119.6.

Pilot: One one nine point six. Good night. I hope I don't talk to you again.

ELM: Okay.

What the pilot had not heard was a conversation just previous to that exchange, between the same controller and one at Ithaca Tower. The tower controller, probably settled in for a long night of inactivity, answered the phone:

ITH: This is a joke? Go ahead.

ELM: This isn't a joke, it's, ah, Cessna 5161R, a Cessna 172. He'll be on the ILS, he should be there at 2225 (5:25 p.m.).

ITH: This guy's going to actually try it, huh?

ELM: Yeah, yeah.

ITH: (Laughter) Okay. Enjoy.

The tower controller confirmed with airport ground vehicles that the runway was clear of traffic, and got a report on the braking action. When the pilot called the tower, he was told:

ITH: Cessna 5161R, Ithaca Tower. Runway 32, wind 320 at five, altimeter 29.99. Runway is reported bare and wet, braking action reported good.

Pilot: Six One Romeo. Does that mean someone actually landed?

ITH: Cessna 61R, negative. Braking action was reported good by a truck. Current Ithaca weather: indefinite ceiling zero, sky obscured, visibility zero, fog.

Pilot: Six One Romeo.

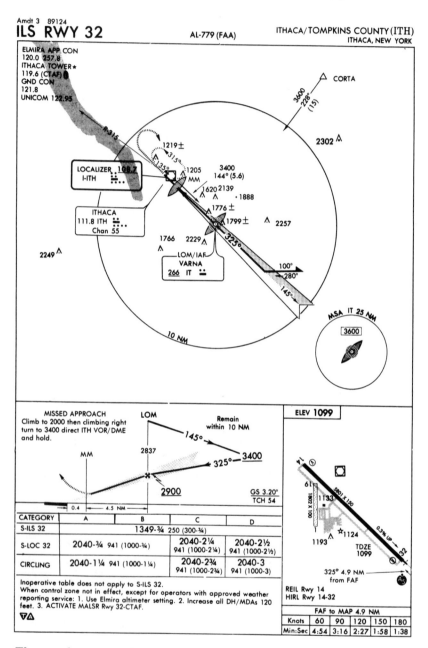

ILS RWY 32   AL-779 (FAA)   ITHACA/TOMPKINS COUNTY (ITH)
ITHACA, NEW YORK

Amdt 3 89124

ELMIRA APP CON
120.0 257.8
ITHACA TOWER ★
119.6 (CTAF)
GND CON
121.8
UNICOM 122.95

MISSED APPROACH
Climb to 2000 then climbing right
turn to 3400 direct ITH VOR/DME
and hold.

ELEV 1099

CATEGORY	A	B	C	D
S-ILS 32		1349-¾ 250 (300-¾)		
S-LOC 32	2040-¾ 941 (1000-¾)		2040-2¼ 941 (1000-2¼)	2040-2½ 941 (1000-2½)
CIRCLING	2040-1¼ 941 (1000-1¼)		2040-2¾ 941 (1000-2¾)	2040-3 941 (1000-3)

Inoperative table does not apply to S-ILS 32.
When control zone not in effect, except for operators with approved weather
reporting service: 1. Use Elmira altimeter setting. 2. Increase all DH/MDAs 120
feet. 3. ACTIVATE MALSR Rwy 32-CTAF.

FAF to MAP 4.9 NM

Knots	60	90	120	150	180
Min:Sec	4:54	3:16	2:27	1:58	1:38

The weather was so bad that initially, the tower controller took the pilot's intention to shoot the approach as a joke. It was no joke, however. The Skyhawk was cleared to land, then descended into trees 2,000 feet short of runway 32.

ITH: Cessna 61R, not in sight, cleared to land, runway 32.

Pilot: Six One R. (Silence for 30 seconds.) Six One Romeo. I assume you've got the, ah, lights up bright?

ITH: Affirmative.

This was the last message from the Skyhawk. It crashed into trees 2,000 feet short of the runway, about 900 feet to the left of centerline. The first tree was 78 feet tall; the wreckage came to rest at about field elevation. The pilot died, his brother and two other passengers were injured.

The brother recalled the flight in his statement to the NTSB, which said in part: "Received approach information and weather; weather at Ithaca was zero-zero. Brother said he would fly ILS and look-see. Passed OM with light and tone. I noted deviations on CDI and GS needle (below GS at least once). Approach not full scale but 1 or 2 dots above. Had started to pick up another beacon, then faded. I think he saw runway lights on right side...then aircraft hit."

Perhaps the Ithaca Tower controller supplied an epitaph when, worried about the extended silence after the last transmission, he called Elmira:

ITH: For the purposes of recording here, ah, verify that you did give this aircraft the weather at Ithaca?

ELM: Yeah, he got it.

"Messy Weather, Misty and Rainy"

In an earlier volume of the *Command Decision Series,* we pointed out the manifold dangers of attempting to maintain VFR flight in IFR conditions. This type of operation is particularly deadly because it involves physiological factors that affect all of us; it doesn't seem to matter if it's a non-instrument rated pilot in a small, poorly equipped airplane or a well-qualified pilot with all the bells and whistles money can buy—trying to stay visual in a low-ceiling, low visibility environment is an invitation to disaster.

The pilot of a Beech King Air found out the hard way while trying to complete a landing from an instrument approach at Mayfield, Kentucky. He was certainly well qualified; an instrument instructor, he had accumulated 6,000 hours of pilot time including 375 in the King Air, and flew the airplane on a regular basis.

The nighttime crash occurred at the end of an IFR flight from Orlando, Florida. The weather at the nearest reporting station—Paducah, 20 miles away—left a lot to be desired: The ceiling was 900 feet overcast, with a scattered layer at 200 feet and visibility three miles in rain. According to witnesses at the crash site, the weather there was worse: They estimated the visibility at only one mile, with a low ceiling. One witness described it as "messy weather, misty and rainy."

The King Air departed Orlando at 7 p.m. and proceeded normally to the vicinity of the Cunningham VOR, at which point Memphis Center issued a clearance for the VOR/DME-A approach into Mayfield. The time was about 9 p.m.

The final approach course is 137 degrees, requiring an arriving pilot to maneuver to land on the single north-south runway. The MDA is 600 feet above ground level.

On the approach, the pilot called Mayfield Unicom for an airport advisory, which indicated a landing to the south; the pilot said he would come over the airport and circle to land on Runway 18. The radio operator then went outside to look for the airplane. He first saw it when it was passing near the airport's rotating beacon, below the clouds and less than 100 feet above the ground.

The radio operator ran back inside to warn the pilot that he was too low, but he never got a response; by the time he got back to the radio, the King Air had crashed.

Another witness was driving near the airport in the direction opposite the flight path of the airplane. He saw it about a half-mile before it crashed, and said he thought it was in a nose-low attitude. Ten seconds later, he saw a fireball as the Beech struck the ground.

The crash site was about a mile and a half northeast of the airport, in a position that would have corresponded to a left base leg for Runway 18. The airplane impacted a large oak tree about 15 feet above the ground, at which point the left wing was severed and the landing gear was ripped off. Bits of wreckage were strewn along a 700-foot path from the tree to the fuselage and the remainder of the airplane, most of which was consumed by fire.

Examination of the wreckage failed to disclose any malfunction or mechanical failure that might have led to the crash. The wreckage configuration and apparent shallow descent angle made it unlikely that the pilot stalled the airplane; but trying to bend a King Air onto the final approach in very low visibility and with only 100 feet or so of ceiling is a very delicate task.

"There oughta be a law..." And there is. The portion of FAR 91 that applies to all instrument pilots when maneuvering at the end of an approach procedure says that a pilot may not land unless the visibility is at or above the published value (no circling approach—one with an "A" in the title—has a visibility minimum less than one mile). The regulation also states that when a descent is commenced from MDA, the airplane must *continuously* be in a position from which a landing can be made using normal procedures and normal rates of descent.

As we said, fly every approach as published and follow the rules; you won't hit anything.

IFR Short Cuts:
Sometimes Good, Sometimes Very Bad

When an instrument pilot candidate is cramming for the written test, he knows literally how to read an approach chart upside down, and can recite chapter and verse about limits of IFR airways, tolerances for VOR receivers, and the like. Entries into holding patterns and initial approach procedures are part of this arcane knowledge.

On the eve of the exam, many IFR pilots-to-be will have etched in their memory banks a picture of the recommended procedure for entering a holding pattern. But the seasoned instrument pilot may find that controllers generally anticipate the most expedient entry into a holding pattern and give vectors that often lead right onto the final approach course, without the need for complicated entries and procedure turns. And if a procedure is unclear, the canny pilot will confess that he doesn't understand what the controller expects and—even at the risk of appearing a little slow-witted—will ask questions until he does.

An unasked question of this kind appears to have played a part in the crash of a Beech Baron 58 at Gainesville, Florida. In an accident compounded by other faulty IFR procedures, the pilot and his two passengers died when the plane struck the guy wires of a cable television antenna west of Gainesville Airport during an attempted ILS Runway 28 approach.

The pilot held a commercial license and had 3,100 total hours, including 1,168 in the Baron. He held an instrument rating and was presumed to be experienced, but investigators were not able to find the logbooks after the accident, and weren't able to detail his IFR experience.

The Baron departed an airport in West Virginia and was originally headed for Citria, Florida. The trip had begun without a flight plan, and no radio communication took place until a point in southern Georgia, when the pilot requested VFR radar advisories.

As the flight got into northern Florida, the pilot obtained a weather briefing from a nearby FSS and was informed that his destination airport was experiencing IFR conditions. The airport serving Citria had no instrument approach procedure, so the pilot requested an IFR clearance to Gainesville, where the weather was above minimums— but not much.

He was cleared to the Taylor VOR at 9,000 feet, then south on Victor 157 to Gainesville. Passing Taylor, the pilot was cleared direct to WYNDS, the LOM for the ILS Runway 28 approach at Gainesville. The controller offered a heading to fly "until able to navigate direct," but the pilot promptly announced he was picking up a steady signal from the WYNDS NDB and could fly direct.

The Baron—N4158S—was cleared to descend to 7,000 feet, and a few minutes later to 4,000 feet.

The controller had another Baron inbound for the same approach at Gainesville, and he coordinated with Gainesville Tower as he estimated their arrival times. He had earlier thought the other aircraft would arrive at WYNDS first, but now he saw a way to expedite the situation by letting the 58S go first. Within a minute and a half after the clearance down to 4,000 feet, the controller radioed: "Four Five One Eight Sierra, can you give me a little better rate of descent? I'll be able to fit you in first." The pilot replied, "Five Eight Sugar, roger, we'll drop it out."

And he did, with a descent rate of 1300 feet per minute, based on his level-off report time. The pilot's effort was rewarded with a clearance for the ILS 28 approach as the

The Baron pilot's position awareness went rapidly downhill as the approach into Gainesville progressed. At one point, the pilot passed over WYNDS going in wrong direction. The aircraft eventually struck antenna guy wires, killing all aboard.

Baron passed over the WYNDS radio beacon southeast bound. He was handed off to Gainesville Tower, and this conversation followed:

Pilot: Gainesville Tower, Baron Four One Five Eight Sierra, with you—ah, ah, picking up the localizer.

Controller: Baron Four One Five Eight Sierra, Gainesville Tower. Report the WYNDS outer compass locator inbound. [A reasonable request; remember that the Baron was headed southeast at the time.]

Pilot: Five Eight Sierra, say again?

Controller: Baron Four One Five Eight Sierra, report the WYNDS outer compass locator inbound.

Pilot: Five Eight Sugar.

This was probably the pilot's most opportune time to sort out any lack of clarity in what he was expected to do, but the pilot acknowledged the instruction and let it pass. Recorded radar data showed that he flew over the WYNDS beacon and began a right turn. About two minutes later, this conversation:

Pilot: Five Eight Sierra, ah, at the outer marker.

Controller: Baron Four One Five Eight Sierra, cleared to land Runway 28, wind is 030 at 5.

Pilot: Five Eight Sugar.

As the radar data retrieval later showed, the plane was nowhere near the outer marker when the pilot reported; the Baron was more likely over the middle marker. This is a very serious error, and was probably the result of an attempt to fly the ILS without a procedure turn or set-up maneuvers of any kind. And as the misguided, unstabilized attempt progressed, it got worse; the last error was the pilot's failure to pull up and make a missed approach.

The plane had gone south of the localizer after passing WYNDS, crossed north of the localizer after the last transmission, and then headed west for the next couple of minutes. Ground witnesses in the area heard and saw the plane go by at low altitude, followed by the sound of a crash. The Baron struck the guy wires of a 670-foot antenna, one of several in an antenna farm located six miles west of the field. The point of impact was 150-200 feet above ground level. The wreckage then fell to the ground and burst into flames, obliterating most cockpit evidence, but it appeared to investigators that both propellers were turning at impact, the landing gear was down and the flaps were partially extended. There was no evidence of pre-impact malfunction of the airplane.

Decision Height for this approach was 200 feet AGL. The weather at the time of the crash was indefinite ceiling 400 feet, sky obscured, visibility one mile in fog. This was the same as the report given to the pilot before he opted to make the approach.

It appears likely that he got to an altitude where some visual contact with the ground was achieved, and that the pilot was attempting to stay underneath the ceiling and find the airport. The rules require that a missed approach be executed immediately if the runway environment is not clearly visible at DH. With no reliable distance information available, the pilot was probably still looking for the airport when he hit the antenna.

But the crucial, flawed decision was the one that led him to attempt the approach from the LOM in the first place. Why did he do this? No one can answer for the pilot of course, but several possibilities come to mind.

First, when he contacted Gainesville Tower and was instructed to "report WYNDS outer compass locator inbound," the pilot may have simply misunderstood, and believed he was being asked to turn inbound immediately. This we consider unlikely, because such an instruction is usually interpreted correctly.

Perhaps what threw the pilot was the reference to the beacon as the "outer compass locator," a designation usually reserved for a beacon that is situated alone, not co-located with the outer marker—in which case "locator outer marker" is the commonly used term. Even though the LOM was clearly depicted on the approach plate, he could have become confused. This would have been the ideal time to confess that he didn't understand the clearance, and seek a clarification—as embarrassing as that might be.

The localizer needle may have been moving to center just as the controller called for a report of the marker inbound, inducing the pilot to take a chance. Contrary to the belief of some instrument pilots, it is perfectly legal to perform (or omit, for that matter) any maneuvering the pilot deems necessary, within the rather generous limits on the chart. So it is also possible that the pilot was not confused, but merely taking a rather unsafe "short cut."

Finally, it is possible that thin spots in the clouds gave the pilot the impression that conditions were better than reported, and that strict adherence to the instrument approach procedure would not be necessary.

The outcome of the unfortunate event proved—as it often does—that an orderly transition onto the initial approach and a long, stabilized final approach on localizer and glideslope, is always a safer way to conduct an IFR arrival.

Between a Rock and a Hard Place

Corollary 86 of Murphy's Law states that when things have gone from bad to worse, you ain't seen nothin' yet. This is particularly true when trying to salvage one of those not-quite-perfect VOR approaches by circling to land, as an experienced air taxi pilot found out one night at Brainard Field in Hartford, Connecticut.

The 5,800-hour commercial pilot held multi-engine and instrument ratings and had been busy, to say the least. The day before the accident, he had flown an air taxi trip that started at 9:30 p.m. and lasted until 7:30 a.m. the following morning. He went off duty, but his sleep was interrupted several times by phone calls; he conceded that he was fatigued when he reported for duty at 7:30 that evening. It was 11:30 p.m. when he started his homebound leg from Providence, Rhode Island, to Brainard.

The trip began VFR, but he air-filed en route, since the airport was reporting a ceiling of 1,000 feet overcast, visibility three miles in rain and haze, wind was out of the northeast at five to eight knots.

The Piper Aztec was vectored onto final and cleared for the VOR-A approach; the pilot contacted the tower and was advised that he should expect to fly over the airport, then enter a left traffic pattern for Runway 2—another aircraft was on left base for the same runway. The controller said he later offered the option of a left base to Runway 2. The VOR approach final course is 334 degrees, or about 45 degrees off the runway centerline.

The pilot told investigators that he missed the first VOR approach and was given vectors for a second try. He reached the MDA (581 feet AGL) and saw Runway 2, but was "in no position to make a safe landing—too high and too close." The controller said the pilot requested the runway lights to be set at the highest level, whereupon he reported the field in sight.

After realizing he couldn't make Runway 2, the pilot decided to turn left and enter a downwind for Runway 20, and his request was approved by the controller. Still at the MDA, the pilot said he went through a heavy rain shower, then overshot final for Runway 20, partly due to excess airspeed. Nonetheless, he attempted the landing, touching down beyond the mid-point of the 4,418-foot runway. It is fair to say that it would take an extremely skillful pilot with undiminished abilities to stop an Aztec in 2,000 feet of wet runway with an eight-knot tailwind. In addition, this touchdown was probably faster than normal, the flaps were only halfway extended, and despite the adrenaline that doubtless was now flowing, the pilot was fatigued.

Things were bad enough, but now Corollary 86 entered the picture. Realizing that he couldn't stop on the runway, but that it was too late for a go-around, the pilot said he attempted to pull the mixture controls back to idle cutoff; unfortunately, he pulled the propeller controls instead, putting both props into feather. Had he any hopes of a gentle arrival, this destroyed them completely. The Aztec went off the end of the runway, through a chain-link fence and came to rest about 50 feet later. It had suffered about $10,000 worth of damage, but the pilot emerged uninjured.

"Plan Your Flight, and Fly Your Plan" applies to just about every aviation procedure imaginable, and especially to an instrument approach that starts to go sour. A judicious missed approach is often the better thing to do; if nothing else, you can return for another try with a lot of information you didn't have the first time.

Little Things Mean a Lot

Instrument pilots place near-total reliance on their altimeters during the execution of approach procedures. Except for most ILS approaches, in which the coincidence of the middle marker and Decision Height on the glide slope is the norm, the altimeter provides the only meaningful information about the distance between airplane and ground; but this information is only valid when the altimeter is properly adjusted.

The insertion of a correct altimeter setting is important to virtually every approach in IMC, but becomes absolutely critical when the airplane must be flown right down to the minimum altitude before achieving visual contact. And if the pilot elects to depart the published procedure and remain at Decision Height while he looks for the airport, an improperly set altimeter often forges the final link in the chain of events leading up to an accident.

Two instrument-rated private pilots came to an untimely end when their Cessna 210 crashed near the Central Wisconsin Airport in Mosinee, Wisconsin. The role of the altimeter in this crash is too significant to overlook.

The flight had originated in Colorado Springs, with stops at Sioux City, Iowa, and Mosinee. The only weather briefing that showed up in records of the flight was the one obtained from the Denver FSS before departure.

The Centurion departed Sioux City on an instrument clearance, but the pilots canceled IFR about 11 minutes after takeoff. It was next heard from at 8:23 p.m., when one of the pilots contacted the Wausau, Wisconsin, FSS and reported a position west of there, on top at 5,500 feet. In consideration of the current weather at Mosinee (indefinite ceiling 500 feet, sky obscured, visibility one-half mile in drizzle and fog) the pilots requested an IFR clearance.

The Flight Service specialist didn't have the authority to issue the clearance; he advised the pilots to contact Chicago Center, and in due time, the Cessna was vectored onto the localizer, and was cleared for the ILS Runway 8 approach at 8:43 p.m. However, the radar coverage does not allow aircraft to be tracked below 6,000 feet, so the controller could not monitor the progress of the approach.

Shortly thereafter, the pilots declared a missed approach. At 8:55 p.m., again at 9:18 p.m., and again at 9:57 p.m., it was cleared for successive approaches after missing each previous attempt. It was not heard from again after commencing the fourth approach, and because of heavy fog, the Centurion was not found until the next day, 1.4 miles east of the airport at an elevation of 1,277 feet.

The last weather given to the pilots was during the 9:57 communication, after an official observer on the field had reported an indefinite ceiling of 300 feet, sky obscured, visibility one-half mile in fog and light drizzle, altimeter 29.97.

There was no clear evidence of pre-impact malfunction of the airplane. Although propeller damage was consistent with rotation at impact, it had the appearance of a low-thrust condition. Firemen who arrived first on the scene described fuel pouring from the wing tank outlets, but parts of the fuel system were devoid of fuel when the investigator arrived.

The plane's altimeter was found in the wreckage and although an altitude reading could not be obtained, the adjustment window displayed a setting of 29.81. The difference between this and the correct setting (29.97) would cause a pilot to fly 160 feet lower than he intends. The DH for the ILS approach was 1,474 feet. Simple arithmetic shows that these pilots were probably flying within one wingspan of the ground, a hazardous proximity anywhere, but doubly so in darkness and fog a mile-and-a-half from the airport.

Burning One's Bridges

The Federal Aviation Regulations are arranged in three levels, if you will, with regard to standards of care; Part 91 takes a very liberal approach, giving not-for-hire operators a very long leash— "y'all be careful, y'hear?" Move up to air taxis and the like—Part 135—and pilots are required to exercise more concern for airplane capabilities, weather conditions, and so on. When you get to the top of the regulatory heap—scheduled airline operations under Part 121—passengers are entitled to the highest possible standard of care, and the rules tighten accordingly.

These "non-91" rules literally force commercial operators to get more weather information before taking off, to have alternates that are even more "solid gold" than those required for private pilots, and prohibit even the commencement of IFR approach procedures unless there's a weather observer on the airport who reports at least the minimum visibility. When a commercial operator ignores these built-in safeguards, the result is often an unhappy one.

For example, the pilot of a Cessna Caravan suffered serious injuries when his airplane crashed into trees short of the runway at Shiloh Airport in Madison, North Carolina. The Part 135 cargo flight—filed IFR all the way—departed the Greater Cincinnati Airport at 6 a.m. bound for Greensboro, North Carolina, with an intermediate stop at Roanoke, Virginia.

After concluding his business at Roanoke, he was off again, and contacted Greensboro Tower when he was about 50 miles north of the city. He told the controller that he had the current ATIS information. The controller responded by issuing the current Greensboro altimeter and advised that the RVR (runway visual range) was 1,600 feet—short of this pilot's requirement of RVR 1,800.

Part 135 contains no provision for a "look-see," so the Caravan pilot requested the weather at Winston-Salem Airport. Not good enough there, either; a 300-foot ceiling, sky obscured, with one-quarter-mile visibility in fog. The pilot elected to hold, and was cleared to do so.

Within a short time, it appeared that things weren't going to get any better; Winston-Salem had dropped to zero-zero, and the pilot told Greensboro Approach Control that he wanted to go to Shiloh Airport. He said he would try a visual approach at Shiloh, since the weather there was good when he flew over it on his way to Greensboro. If the visual approach wasn't possible, he'd try the NDB approach. This plan was approved.

Following the Cessna Caravan on radar, Greensboro Approach told the pilot when he was over Shiloh. Alas, that bridge had burned as well—the pilot said he couldn't see the airport and he requested the SDF (Simplified Directional Facility) approach. The controller vectored him onto the final approach course, and subsequently cleared him for the approach.

The Caravan crashed into trees 1.8 miles from the runway threshold. The missed approach point for the SDF approach is one mile from the threshold, and the minimum descent altitude is 1,120 feet. The pilot later told investigators that his altimeter was reading 700 feet after the airplane crashed. Weather conditions at Greensboro at the time of the accident included a ceiling of 100 feet, sky obscured, with slightly more than one-tenth mile visibility.

There's a lesson here for all not-for-hire pilots. If these extra-conservative precautions are required for professional pilots (often two of them on the flight deck) with lots of experience, lots of training and lots of help from better instruments and more of them, they should at least be considered by single pilots flying IFR in lower-performance, lesser equipped airplanes. "Y'all be careful" must not be taken lightly. Being permitted to take more chances is not a mandate to do so.

Lured by the Muses of the Night

One of the big problems associated with misunderstood and misin-terpreted radio communications is the fact that we often hear what we *want* to hear, or what we *think* we hear. The same is apparently true in the visual realm, because there are numerous accidents in which pilots mistake lights or other visual clues when everything else at their disposal makes it clear that what they are seeing couldn't possibly be what they're looking for.

How many times have you been so intent on finding a strange airport on a hazy day that you lock onto the first one you see, only to be embarrassed when it turns out to be some other field? And frequently, this in spite of the fact that you're still several minutes away from the real destination, or that your navigation displays tell you clearly you're nowhere close.

Transitioning from the total lack of outside visual references that is part and parcel of IFR operations to the sudden appearance of ground lights during a night approach has been the undoing of many instrument pilots. Witness the problems encountered by this Cheyenne driver. The pilot was killed and his three passengers suffered serious injuries when their Cheyenne struck powerlines, trees and then a house while attempting an ILS approach to Lambert Field in St. Louis, Missouri.

The business flight had departed Alexander City, Alabama, on an IFR flight originally intended for Quincy, Illinois, but diverted to Lambert due to the weather at Quincy. Weather at Lambert was also poor, including a sky partially ob-scured, an overcast at 100 feet, visibility three-eighths of a mile in fog and rain.

With routine handling and no evidence of a problem in any of the pilot's transmissions, the aircraft was vectored to intercept the final approach course of the ILS Runway 12L approach. Whereas most hand-flown intercepts involve at least a few bracketing maneuvers, radar showed the Chey-enne merging smoothly onto the localizer course—evidence consistent with the use of an autopilot.

The plane followed the localizer course accurately until within about half a mile of the airport, when it deviated slightly to the left. At the same time, the data shows the plane was consistently below the glideslope throughout the approach, and that its airspeed varied from time to time. The plane struck powerlines, flew through a stand of trees, and crashed into a house.

The NTSB investigator received reports that there had been prior problems with the Cheyenne's flight director, which could have had a bearing on the vertical excursions during the approach.

But more to the point we're trying to make, the investigator observed that a lighted cargo area is situated to the left of the runway, and that the heading deviation began at a point consistent with the pilot's sighting this area instead of the runway. Further, the pilot's son reported that his father's consistent habit was to wait to drop the landing gear and flaps until the runway was in sight. The surviving passengers told investigators they had heard the rumble of gear extension, saw some lights, followed instantly by the initial impact.

"Get-Home-Itis" Strikes Again

Four o'clock in the morning is not one of the prime times to be flying, especially when you're contending with IFR conditions at the end of a long trip that's been one disappointment and delay after another. The strong urge to get home apparently affected a Mooney pilot as he flew an early morning ILS approach at Peoria, Illinois; unfortunately for the pilot and his passengers, a power line got in the way.

The Mooney had been part of a group of several airplanes and pilots from the Illinois area who annually flew to the Bahamas on vacation. As the group prepared to depart Freeport for the trip home, the Mooney pilot discovered a magneto problem which couldn't be repaired in the islands. The rest of the group departed and, when they reached Florida, obtained a magneto and air-expressed it to the stranded pilot in Freeport.

The Mooney's departure time from Freeport was unknown, but the flight to the mainland takes less than an hour. The plane arrived at Fort Pierce, Florida, and departed there at about 8 p.m. EST. The pilot had listed his destination as Huntsville, Alabama, but later amended it as Nashville, Tennessee, for unknown reasons. Flight time to Huntsville is normally about four hours, and the extension to Nashville added about half an hour to the flight. It is noteworthy that the pilot arrived at Nashville low on fuel and expressed concern to the tower about his sequence in the pattern.

The tanks were topped and the Mooney headed for Peoria, arriving in the destination area around 4 a.m. Greater Peoria Airport was reporting sky obscured, ceiling 100 feet, visibility one-quarter mile in fog. The pilot remarked that he would try an approach to Peoria and if that failed, would divert to Rockford, Illinois.

The Mooney was cleared for an ILS Runway 13 approach, but it struck powerlines two and a half miles short of the runway, on the extended centerline. There was no evidence of engine failure or other malfunction of the aircraft prior to impact.

No Gas and No Runway in Sight

Aside from several accounts of rather bizarre system failures, the *Aviation Safety* files were rather bereft of IFR emergencies, except for two areas in which pilots got themselves into trouble while flying IFR fuel-exhaustion/starvation and missed-approach procedures. We relate several of these for your consideration.

"Cleared to Hold"—Join the Crowd

Large airplanes often depart with less than full tanks, because it's more important to carry revenue-producing passengers or cargo. But most light airplane pilots are able to "top it off" without exceeding weight limitations. Especially when the upcoming flight will encounter adverse weather, it behooves a pilot to carry all the fuel possible—just in case. Here's the story of an Archer pilot who should have filled the tanks before launching on the last leg of his trip. Fuel exhaustion was the final link in the accident chain, but as you'll see, there were other contributing factors.

The pilot and his wife had departed Grand Canyon, Arizona, earlier in the day and had stopped at Lancaster, California, for fuel. An in-person weather briefing indicated that VFR flight was not recommended; the pilot neverthe-less left for Napa, California, without filing a flight plan.

Prior to departure, the pilot had requested that line personnel top the fuel tank on one side, but fill the other only to the indicator tab, providing a total endurance of slightly more than four hours.

The pilot had logged 1,123 total hours, including 758 in the Archer, which he owned. He had flown 36 hours within the last 90 days, and 15 hours in the past month. He held an instrument rating and had logged 74 hours of simulated and 12 hours of actual instrument flying. In the past six months he had logged only four hours on instruments, the most recent entry a flight of 2.5 actual IFR hours four months prior to the crash. Likewise, his most recent entry showing a night flight was more than three months earlier.

The flight went without incident until arrival in the Napa area. The pilot contacted Oakland Center at 5:13 p.m., reporting eight miles west of the Scaggs Island VOR at 9,000 feet, and requesting the VOR approach to Napa. The VOR serves as a feeder not only for Napa, but for two other airports. There were multiple layers of clouds extend-ing almost to the ground and numerous instrument ap-proaches were being conducted, with upwards of eight aircraft holding at the VOR. Controllers consequently instructed the Archer pilot to hold where he was because of traffic.

Prior to this instruction, he was asked whether he was in VFR conditions and he responded in the affirmative—but a King Air crew climbing out through the same area reported being in solid IFR conditions, with rime ice form-ing, until reaching 10,000 feet.

Eight minutes after initial contact, the pilot cleared direct to the VOR for a hold at 8,500 VFR, and told to expect a delay of 20 minutes before further descent.

At 5:37 p.m., the pilot was told to descend to 7,000 feet, hold in the standard pattern, and to expect further descent at 5:55 p.m.

At 5:42 p.m., the pilot was given a descent to 5,000 feet for further holding.

At 5:50 p.m., the pilot reminded the controller that he

was "still holding at five." The controller responded that he could expect the approach clearance in about 20 minutes.

Three minutes later, the pilot again stated that he was at 5,000 feet and asked, "How much more delay?" The controller responded that he had one aircraft still holding below the Archer. Three minutes later, the Archer was cleared down to 4,000 feet and six minutes later, to 3,000 feet. The pilot was then vectored outbound on the 230-degree radial, made a procedure turn and started inbound. The flight was cleared for the approach at 6:04 p.m., and the pilot was handed off to the tower.

About a minute later, the pilot contacted the tower and was told to report the VOR inbound, which he did five minutes later, but this wasn't an accurate report. The Archer had actually passed to the east of the airport.

Center controllers who were observing the flight on radar called the tower controllers to warn that the Archer was now four miles east of the airport and headed toward mountains. Tower controllers instructed the pilot to immediately execute the missed approach procedure—a climbing left turn back to the VOR.

The ATC tapes convey a sense of confusion in the cockpit at this time. The pilot said "I have a fire on the right wing," then "I've lost my gyros." Oddly—even though earlier he apparently had not recognized his position relative to the airport—the pilot now gave an accurate DME position. He was immediately started on no-gyro procedures and given a vector directly toward the airport, but contact was lost within seconds.

The airplane impacted in a steep wing-low, nose-down attitude consistent with loss of control. Despite the pilot's statement, no part of the wreckage exhibited any fire or heat damage. Although the propeller was turning at impact, one fuel tank was bone-dry and the other contained about three gallons of fuel.

Even though the pilot may have arrived at Napa with the required fuel reserve, it had been eaten up by the holding time of nearly an hour. The pilot never declared an emergency, or even told controllers that he needed priority service due to a low fuel state.

First Things First

When an instrument approach turns into a missed approach, the initial procedure should be the same no matter what's published on the approach chart—your first priority is to get away from the ground. There are two conditions to be considered: it's one thing to add full power, adjust the pitch attitude, and watch the altimeter and the VSI come to a stop; it's something else to be actually climbing.

Not until the VSI shows a positive rate of climb and the altimeter begins to move upward is it safe to conclude that the airplane is indeed climbing. Attention to other matters, such as retracting landing gear and wing flaps, adjusting climb power, etc. must wait until a climb is definitely established.

An accident which brings this lesson into sharp focus occurred in the early hours of a July morning when a Piper Lance struck the top of a pine tree near Asheville, North Carolina. The four occupants escaped with their lives, probably because the plane didn't hit the trunk of the tree.

The Lance's owner, a private pilot with 524 hours (244 in this airplane) was at the controls. He had been working on an instrument rating, and had logged nearly 15 hours of IFR training under actual conditions in the past 90 days.

In the right seat was his instructor, a CFII with 469 total hours, 73 in Lances. He had flown 114 instrument hours, 73 of them simulated; most of his actual instrument hours had been experienced in the past 90 days. He had been engaged to fly from the owner's private airfield near Charlotte, North Carolina, on a trip eventually destined for a Chicago-area field. Two other passengers were with them at the outset, and another was to be picked up in Asheville.

They set out early, getting off from the owner's strip at 4:50 a.m. They had obtained a weather briefing, and were not surprised at the fog which greeted their pre-dawn arrival at Asheville—the field was reporting a ceiling of 100 feet, sky obscured, visibility one-eighth of a mile in fog and haze, temperature 67, dewpoint 67. Although the visibility was well below published minimums, they elected to try the ILS Runway 34 approach—even though the probability of success was almost zero.

The pilots were vectored onto the ILS final approach course and subsequently cleared for the approach, which they elected to execute with the autopilot. The landing gear was down and the aircraft was slowed to 90 knots; flap position was not reported. The owner was the flying pilot and was monitoring the autopilot, while the CFI was monitoring the left seat pilot and looking for the runway.

As they passed the outer marker inbound, they re-checked and agreed that the Decision Height for the approach was 2,400 feet MSL (field elevation is 2,165 feet) and when they reached that point with no runway in sight, the CFI announced the missed approach. The owner disconnected the autopilot, added power and pulled the yoke back for a climb. The CFI noted the start of a climb and the heading, then called the tower to announce the missed approach. The tower cleared the flight to climb on the runway heading to 5,000 feet (this is the published missed approach procedure).

The owner-pilot noticed that he was slightly off the assigned heading and began a correction; seconds later there was pandemonium. The windshield had been smashed inward and the wind was roaring around the cockpit, causing a maelstrom of loose papers and dust.

The instructor took the controls, recovered from what he recalls as a right bank and began to climb on the runway heading. He told the owner to raise the landing gear, which he did, but that was the extent of his help; the smashed windshield had injured his eyes. The passengers helped him recline his seat and got a coat to protect him from the cold wind.

The Lance climbed sluggishly, and the instructor was concerned about a drop in oil pressure. In the face of trying circumstances, the instructor was able to get the airplane to a safe altitude, continued to navigate and communicate, and upon hearing of several choices for emergency landings, chose Charlotte because he felt it had the best emergency facilities.

Because the wheels had been down when the tree was struck, the pilot suspected the nosegear might be damaged, and his suspicions were confirmed; the nosegear light would not illuminate when the gear was extended, and the CFI landed with the expectation that the nosegear would collapse. It did, but there were no additional injuries as the plane skidded to a stop.

The controller on duty at Asheville said he vectored the Lance onto the approach, followed on his radar what appeared to be a normal approach, heard the pilot declare a missed approach, and lost radar contact over the airport—a normal occurrence. He looked out the tower-cab windows and saw aircraft lights over the runway near midfield, then saw the aircraft veer to the right, fly over the north ramp and off into the fog northbound. Nowhere in his statement does the controller mention seeing the Lance in a climb. He heard the pilot's "mayday!" and shortly after began to pick up the plane on radar again. On the first sweep, the altitude was 2,000 feet (400 feet below DH); on the next it was 2,200.

The facts clearly point to a failure by the pilots to assure that the fundamental requirement of a go-around—positive rate of climb—was attained. The NTSB concluded that the probable causes were the owner's improper IFR operation and the instructor's inadequate supervision of the flight.

We might also suggest that because the approach had been conducted on autopilot, the left seat pilot may not have been prepared for the "feel" of the go-around. When he disengaged the autopilot and added power, he inherited a trim condition that may have seemed as though it would result in a climb. Feeling the nose come up and a slight extra heaviness in the seat, as well as seeing the attitude and altitude instruments change, may have satisfied his unschooled requirements for a go-around while he turned his attention to the heading.

When flying on instruments, beware of feelings—they lie.

All Missed-Approach Procedures Are the Same

Well, at least the initial segment of all approach procedures should be considered identical. On even the most imprecise of non-precision procedures, you will wind up closer to the ground than is generally comfortable when you can't see; therefore, the most important thing to do when the missed-approach decision is made is to reconfigure the airplane for a climb. There's not an approach procedure in the book that doesn't provide room for a few seconds of straight flight while you're getting your ducks in a row; in other words, commencing a climb straight ahead is the thing to do, even though the procedure may call for a climbing turn. Five seconds spent in making the change from descent or level flight should be considered five seconds well spent, especially if an attempt to accomplish everything at once ends in disaster.

There's no one remaining to tell how it really happened, but confusion and disorientation are likely candidates for the probable cause of this missed-approach accident. There was no reason to do anything except fly straight ahead—the published procedure calls for a climb to 700 feet before turning—but the evidence suggests that the pilot attempted or permitted a turn.

The Nantucket Airport was clothed in a not-uncommon fog—after all, Nantucket is an island off the northeast coast, a veritable fog factory. Investigators said weather at the time of the mid-morning accident included a fog layer which began 100 feet above the ground, and topped out at 700 feet.

The pilot of the accident airplane had departed Chester, Connecticut, and made a stop at Providence, Rhode Island, before proceeding to Nantucket Island, where he elected to try the ILS to Runway 24.

Radar coverage in the area does not extend below 1,000 feet, but controllers stated the plane was established on the localizer when it went off the radar scopes. A short while later, the pilot declared the missed approach, but reported no problem.

The missed approach procedure calls for a climb straight ahead to 700 feet, *then* a climbing right turn to 2,000 feet on a heading of 345 degrees.

The Lance was found about a mile north of the airport, having impacted trees on a heading of about 115 degrees. The evidence indicated that the airplane was in a 40-degree right bank with the nose down about 15 degrees when it hit the ground.

If there's ever a time when a pilot must really bear down on basic instrument technique, it's during a missed-approach in actual conditions. High-powered single-engine airplanes are especially prone to significant changes in pitch and yaw when the throttle is opened to commence a climb, and the resulting conflict between seat-of-the-pants and instrument indications must be resolved and overcome.

In this situation, be very diligent in your piloting technique, and don't hesitate to apply one of the oldest rules of successful flying— "Do whatever you must with whatever is available to make the airplane do what you want it to do."

IFR Judgment: The Human Factor [6]

A later volume in the *Command Decisions Series* will cover a wide spectrum of "mind and body" problems that plague pilots. Now, we narrow the focus to human factors in IFR operations, a flying environment in which complete attention to the job at hand is an absolute necessity. Once immersed in clouds, a pilot must make the right decisions, perform the proper procedures, and push distractions into the background. The alternative is not acceptable.

As usual, we will use the medium of research findings and accident reports to illustrate shortcomings suffered by instrument pilots, in hopes that you will learn from the mistakes of others.

When Your Routine Works, Stay With It

There is probably no other mode of transportation in use today in which the operators are as thoroughly trained as are airplane pilots. Even at the lowest levels of aviation, pilots learn from the very beginning that "doing it right" is the only way. The repetitive drills in landing and takeoff, engine failures and other emergency procedures has but one objective—to develop a pilot who will more than likely take the proper action almost reflexively when the real need arises.

Instrument flying is no different in this respect. IFR pilots are trained, for example, to use standard approach procedures and most of them continue to develop their techniques into a "same way every time" operation that can be counted on to deliver when the chips are down. But when a pilot changes the way he flies on instruments, things can come unglued. A small change in his routine for an ILS approach may have been the reason why this pilot failed to lower the landing gear on his Beech Baron during a practice approach.

The pilot had 550 total hours, including 250 hours in multi-engine aircraft (all various Baron models). He told investigators his routine procedure was to drop the gear at the outer marker in order to slow down the Baron and to facilitate capture of the glideslope. During the approach in question, the pilot was brought onto the ILS final approach course outside the outer marker, and intercepted the glideslope at the same time. He told investigators he went through his pre-landing checklist, except that he elected to wait until the marker before lowering the gear. The approach was being conducted with the autopilot coupled.

After passing the marker, he again went through a checklist and again failed to lower the gear. The pilot stated that he saw the red warning light indicating the gear was not down, but "it just didn't register."

"Just before touchdown," he said, "the warning horn went off and, mistakenly construing it as the stall warning, I dipped the nose down slightly and advanced the throttles and set the plane down, very smoothly...and unfortunately, onto the tips of the props."

In the space on the accident form where the pilot is invited to recommend ways to prevent similar accidents, the pilot wrote: "This was a pure and simple case of pilot error, for which I take full responsibility. I should have noted the higher-than-usual airspeed down the glideslope, and the meaning of the warning horn. The only reason I can reconstruct was that the interception of the glideslope was not at the outer marker."

Embarrassing to say the least, and expensive too. But there are two extremely valuable lessons here: First, no matter how many times you make the same approach in the same airplane, VFR or IFR, force yourself to run through at least a GUMPFS-type checklist shortly before landing (that's Gas, Undercarriage, Mixture, Props, Flaps, Seatbelts) to be absolutely certain that the critical items have been taken care of; second, whenever an ATC vector onto an approach course smacks of an intercept that won't give you time to get stabilized and properly lined up, turn it down!

How to Hurry into Trouble

It's often a matter of personal pride, and sometimes a bit of grandstanding when the pilot of a small airplane acquiesces to an ATC request for a high-speed approach. This is a good thing to do when it can prevent traffic tie-ups, but when the demands and distractions of flying faster than usual are more than the pilot can handle, "keep up your speed on final" can turn an otherwise ordinary procedure into a mine field.

The pilot of a Beech Duke should have been able to conform to such a request during a night ILS approach at Indianapolis, but fatigue and distractions apparently overcame his ability to fly fast and still get everything done right.

The accident occurred at the end of a long, fatiguing day in the saddle. The pilot was well qualified—5,120 hours, with over 2,200 in the Duke and 722 hours at night.

Upon reaching the terminal area the Duke pilot was advised to expect the ILS Runway 22R approach, and that he would be sequenced ahead of an inbound jetliner. He therefore planned a shallow, fast approach, intending to cross the outer marker at 170 knots, decelerating to 120 to cross the middle marker, then slowing for touchdown as close to the numbers as possible. He advised ATC of this, and was told to maintain his speed as planned.

Inside the middle marker, and about 1,000 feet from the runway threshold, the pilot felt a slight bump that he couldn't account for. He thought nothing of it, believing it to be nothing more than low-level turbulence. The only thing he noticed was that the approach lights seemed "unusually bright." (The pilot later said that the glideslope receiver in the Duke was not reliable close to the ground, routinely indicating a full-scale "fly up" indication when within 150 to 200 feet of the ground.)

Then he felt two more bumps, harder than the first, and knew that he had hit something, but thought it might have been a truck parked at the end of the runway. The pilot made a normal landing and taxied to the FBO.

It was dark and he was tired, so he did not inspect the airplane—that's tired! Instead, he asked the linemen to tie it down and went home. The only damage he noted at the time was that the retractable step was loose.

The pilot said he intended to return the next day to examine the airplane, but the tower controllers beat him to it; they managed to track him down a couple of hours later, somewhat miffed that he had taken out a total of 17 approach lights. (The lights are mounted five to a stand, and the pilot destroyed three stands' worth, plus two more lights.)

Damage to the aircraft was substantial, but amazingly the landing gear was untouched, and the only damage to the props was a slight graze on one side. The bottom of the fuselage and flaps were badly damaged, and one of the main gear doors was bent.

To Go, or Not to Go—That's the Question

It's the first, and often the toughest decision on every flight; far too many pilots make the wrong choice, and wind up injuring or killing not only themselves but also their helpless passengers. A bad decision is almost always a human factors problem, because the warning signals are usually many and obvious.

Researchers at NASA's Ames Research Center took a close look at this critical decision-making process. Figuring that at least in this regard "pilots are pilots are pilots," they studied 59 accidents involving emergency medical service (EMS) helicopters to find out why these pilots made the wrong decisions.

One of their findings that is indisputably applicable to general aviation flights, is that "poor weather conditions are the greatest single hazard." As every pilot knows, the weather rules are highly subject to interpretation, and grey skies create a grey zone within which a pilot can choose to launch if the spirit moves him. The military, realizing that "can do" pilot attitude could combine with bad weather to yield unacceptably dangerous flights, has developed a number of independent risk-assessment programs, and the NASA researchers adopted these to create a computerized system that gives its view of how risky a particular launch will be. The factors taken into account might well be considered by any pilot when making a go/no-go decision.

There were five types of factors; mission, crew, organizational,

Although he took out 17 approach lights, the Duke pilot managed to land the aircraft and taxi into the ramp. He discovered the damage only after tower controllers tracked him down several hours later.

environmental, and aircraft, and up to eight variables for each factor. For example, under "crew" the variables included EMS experience and IFR currency of the pilot, crew rest, the pilot's familiarity with the area en route and at the site, the pilot's familiarity with the aircraft, the number of pilots and aircraft in the operation, the number of hours since the last meal, and pilot credentials.

The system that emerged has been dubbed Safety Assessment for Flight Evacuation—SAFE. It's the researchers intent to take the initial model and have it continue to evolve based on accident statistics and other data that continue to flow in. Their goal is a model that will consider all the relevant factors, then generate an assessment that will wave a big caution flag when risk rises above a certain level.

In the end, of course, the decision will always be the pilot's. But there's much grim evidence that pilots would benefit greatly from having an objective third party available to give its assessment of how things stack up. Some day, a preflight call to a flight service station might well consist of a preflight risk assessment along with the weather briefing.

The Fine Art of Dodging Bullets

With instrument flying, there is no such thing as a mistake too dumb to make. Most instrument pilots can relate at least one confidential story of a simple, bald screw-up that, had it not been caught, could have spelled disaster. The story will end with, "I was amazed that I could be so stupid."

Instrument flying demands so many detailed judgments and actions, each of which can lay the groundwork for an accident, that we believe the simple and too-obvious mistakes are the ones a pilot is likely to make. Gone are the days (we hope) when a pilot would fly into the ground with the ILS needles centered and rock steady, because the unit had failed. This kind of mistake has been replaced with others of the same ilk, however; witness the experience of a pilot who was setting up for a tough approach at Norfolk, Virginia in his Beech Bonanza.

> He had already made a diversion due to bad weather, and Norfolk was reporting 200 feet and 3/16 of a mile. Vectors got him onto the ILS course and he trimmed for 120 mph, lowered the gear and opted to let the autopilot fly the approach. He punched the approach button, and within seconds found himself pitched over to the right in what he described as a "spin." The airplane broke out in time for the pilot to right it, but too low for recovery; it stalled into a swamp and, fortunately, all aboard survived.
>
> NTSB investigators found that the pilot had tuned 109.15 MHz on his ILS receiver, when the proper frequency was 109.10. Although the manufacturer of the autopilot did not believe the mistuning should have allowed the unit to couple to a spurious signal, the NTSB believes this was the cause of the crash.

Not so fortunate were the occupants of a Cessna Hawk XP which crashed near Dover Plains, New York. The plane struck a ridge 12 miles out while the pilot was attempting a VOR approach to Dutchess County Airport at Poughkeepsie, New York. His logbooks were destroyed, so investigators were unable to determine the pilot's flight time, although he was properly rated and had achieved commercial certification about four months earlier.

The flight from Bedford, Massachusetts, went without incident until it was handed off to Dutchess Tower—then a non-radar facility—for the descent and approach.

The chart for the VOR-A approach depicts the Kingston VOR as the initial and final approach fix; it lies about 3-1/2 miles from Runway 6. It also depicts the Pawling VOR, which is likewise off the approach end of Runway 6, but about 15 miles away.

The MDA is 760 MSL, field elevation 165, and weather that night was estimated at 1,800 overcast, three miles in haze, rain and fog.

The approach calls for crossing the VOR and turning outbound, then a procedure turn and inbound. However, upon initial contact, the controller told the pilot to hold over the Kingston VOR and expect an approach clearance in about 12 minutes. This would involve crossing the navaid and making right turns in the holding pattern.

Less than two minutes later, the controller cleared the pilot for the approach. Two minutes after that, the controller asked whether the plane was on the outbound leg of the approach, and this exchange took place:

Pilot: Ah, no. We're, ah, completing the first holding pattern.

Controller: Why did you make a holding pattern? I don't understand.

Pilot: We were told to do so by the controller.

Controller: No, sir, you were cleared about, ah, four minutes ago for the VOR Alpha approach.

Pilot: Roger, sir. Well, we're not quite back onto the VOR now, level at four.

Controller: Roger, sir. Report the VOR outbound, please.

This the pilot proceeded to do, and shortly reported inbound at 3,200 feet, then crossing the VOR, at which time he was cleared to land. The controller was probably trying to determine why the airplane was not in sight when it crashed.

In a nutshell, the pilot had mistuned his nav radio and executed the entire approach using the Pawling VOR. The crash site was in the precisely right position for a descent to a 1,250-foot ridge, not an airport.

A related story with a happier ending came recently through the NASA Aviation Safety Reporting System newsletter, *Callback*. Reports are de-identified under the system, but the essential mistake being admitted by the pilot is a classic:

"The conscientious and alert Center controller was still observing the aircraft on radar after handing it off to the non-radar-equipped Approach Control. The airplane was vectored to intercept for the LOC-DME-B (back course) approach, maintain 10,000. When the aircraft was established on the radial, Center instructed it to go to Approach. Approach instructed the aircraft to maintain 10,000 until 21 DME, then cleared for the back course approach.

"Everything trimmed up, tracking inbound right on the button, autopilot switched to ILS (and ILS DME), pilot verified he had a few miles to go at 10,000, flipped the autopilot on, and blithely went about cleaning up the cockpit (putting away maps, other approach plates, etc.). He diverted his attention for at least a full minute. When he next glanced at the DME readout, he was just inside the 21-mile checkpoint, and began a profile descent.

"What he failed to observe is that he had placed the autopilot in the 'normal' tracking mode instead of the 'reverse,' or 'back course' mode, and while his attention was diverted, the autopilot took the aircraft through a beautiful, coordinated 180-degree turn. He was now heading due wrong!"

The Center controller noted the course reversal on radar and phoned the Approach controller, who contacted the pilot, verified the incorrect heading, then requested in a very calm voice "Give me your best rate of climb immediately." The back course approach is characterized by high terrain very close to the descent area.

The pilot was taught a great lesson—never, *never* divert your attention from the instruments once you are potentially below terrain on an approach. This is good advice, but we would go farther;

what an instrument pilot needs to remember is that when things are going unbelievably well, there's probably a reason.

A high-time commercial pilot was killed and his passenger seriously injured when their Cessna 425 turboprop twin struck the ground during a missed approach at Sanford, Florida.

After an 8:20 p.m. takeoff from Atlanta, the Corsair was arriving in the destination area shortly after 10 p.m. when the pilot was informed by Orlando Approach Control that the Sanford tower was closed, but had reported an hour earlier that fog was forming. The only official observation was that at Orlando, which reported a 3,000-foot broken ceiling, visibility 10 miles. The Cessna was cleared for the ILS Runway 9 approach at Sanford.

According to the passenger, a 500-hour private pilot, the approach was normal until the Corsair was low to the ground, when the runway lights were observed through patchy fog, but off to one side. The pilot was advancing power for a go-around when the plane struck the ground. Wreckage evidence indicated the landing gear was up and flaps were extended when the plane struck between the runway and a taxiway, bounced across the taxiway and came to rest about 750 feet beyond the initial impact point.

When the flight failed to close its IFR flight plan, a ramp check was conducted, but the crashed Cessna could not be seen because of the fog. It was discovered the next morning by maintenance personnel who had been dispatched to that part of the airport. The plane's emergency locator transmitter had activated, but its antenna had broken off, limiting the signal reception distance to only a few feet.

A medical examiner told investigators the pilot died of head injuries during the impact sequence—no surprises there, because the pilot was wearing a seatbelt but not a shoulder harness. If that weren't enough negligence in the restraint-system department, a bicycle was being carried in the cabin, partially secured by the shoulder harness of the copilot's seat. When the aircraft came to rest, the bicycle jammed the cockpit area, preventing the passenger from exiting. He was extricated by rescue personnel the next morning.

According to preliminary medical results, the pilot's body contained the drug dextromethorphan, commonly found in over-the-counter cough medicines; the amount apparently used by this pilot was more than sufficient to cause drowsiness—and at a time when he needed the unrestricted use of all his faculties.

The Limits of the Envelope

Professional test pilots are paid well to take chances, and perhaps they really earn their keep when they make the no-go decision. Before they undertake to find out just how fast an airplane will go or how it will recover from an unusual inflight situation, they will carefully assess the risks and alter the program if it appears that the risks are greater than any possible benefit.

If more non-pro pilots adopted the same attitude (if for no other reason than they are *paying* to fly, quite the opposite of the test pilot!) the accident rate might decline somewhat. When an airplane's limits have already been determined, why press your luck? A Beech Travel Air pilot will not be able to relate his reasons for trying to squeeze every bit of range from the airplane—he and his wife were killed when the light twin struck trees and then the ground during an attempted localizer approach to Runway 10 at Millville, New Jersey. Although the pilot had declared a missed approach, the plane crashed about a mile and a half west of the airport, without ever reaching the field.

The NTSB report indicated that logbook entries and other records pertaining to the pilot's experience were sketchy, but it appeared he had more than 400 total hours and at least 100 hours in the Travel Air, which he had owned for several years.

Upon arriving in the southern New Jersey, the pilot was advised low ceilings and visibilities prevailed throughout the area. Dover Air Force Base reported a ceiling of 200 feet with the sky obscured, and Millville reported an estimated 400-foot broken, 600 overcast, visibility one mile in light drizzle and fog. However, at the time of the 6:50 p.m. crash, the Millville FSS had made a special observation which indicated the ceiling at 300 feet, sky obscured, visibility one-quarter mile in drizzle and fog, temperature and dew point both 55 degrees.

The pilot was vectored onto the localizer final approach

course without incident and was instructed to descend, eventually reaching as low as 1,100 feet. The LOC Runway 10 approach uses a marker as the final approach fix , and the pilot reported passing the marker inbound at 6:46 p.m. At 6:49 p.m., the pilot told the FSS (no tower at Millville) that he had sighted some lights, but then said he was making a missed approach. The FSS staffer acknowledged and called approach controllers to report the missed approach, but the pilot was not heard from again.

At the same time, an FSS specialist who had gone outside to take the weather observation noted that no airplane was heard or seen flying over the airport. However, residents west of the airport reported hearing a low-flying plane, and one mentioned that he thought he heard it turn toward the south before he could no longer hear the engine sound.

The plane was found west of the airport, having struck trees in a slightly left-wing-low attitude while on a heading of about 112 degrees. Although physical evidence indicated no pre-impact malfunction of the airplane and substantial power on both engines at impact, investigators were attentive to the question of the fuel status, since it may have had a bearing on the pilot's attempt to land at Millville despite the poor weather.

The flight plan estimated a cruise speed of 165 knots, time en route five hours and ten minutes, and six hours and ten minutes of fuel on board. The FBO at Sarasota told investigators the pilot had been careful to make sure the plane was fully topped off for the flight, even to the point of asking for two more gallons of fuel to make the tanks absolutely full. The FBO also said the pilot told him he normally stopped to refuel on the trip to New Jersey, and this was the first time he would attempt the trip non-stop.

The pilot had obtained his private certificate in 1959, and multi-engine and instrument ratings in 1981. Acquaintances told investigators that the pilot often filed IFR on VFR days, but disliked flying into actual IFR conditions.

FAA records revealed that the pilot had once survived a crash in which he had flown into clouds and turbulence associated with thunderstorms and struck a ridge while attempting to descend into visual conditions. In the wake of that prior accident, the FAA had

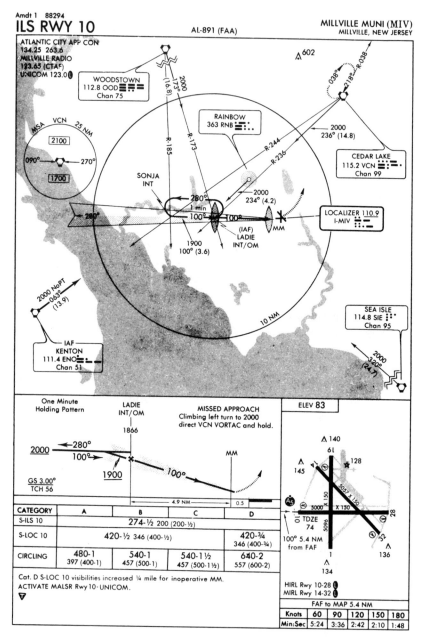

*The Travel Air appeared to intercept Millville's localizer normally,
yet it never completed the approach. Wreckage was found west of the
airport, where the aircraft struck trees.*

suspended the pilot's license for three months.

There's a wealth of human factors to be considered in this accident, not the least of which was the conscious, planned attempt to stretch the airplane's range to the maximum. There's nothing basically wrong with such a plan, nor is it inherently dangerous, as long as there are adequate alternates available to accommodate the slight errors in fuel burn that can become critical over a period of five hours or more. In this case, the pilot had put himself in an unforgiving situation because of the weather conditions in the destination area. If you're going to push the limits, take a lesson from the test pilots; do it under ideal, controlled conditions.

Taking the Measure of Complexity

Saying is believing—or at least it was for investigators who asked a group of pilots about what was going on in their heads as they flew an IFR cross-country flight in a simulator. The goal was to tease out the underlying model of reality that allows a pilot to translate the rules of flying into actions that enable him to fly from hither to yon.

There were two groups of pilots for the experiment; one batch consisted of experienced pilots who either flew commercially or were IFR instructors, the others were private pilots with less than 100 hours of flying time. One of the purposes for including novices was to see if the strategies pilots use for solving problems is affected by experience, or whether people continue to fly as they were taught.

Each pilot was given a battery of psychological tests, followed by simulator training, then the experimental flights.

The task was basically that which faces every pilot on every flight—to fly the airplane without incident, coping with any irregularities or difficulties as they arise in the best way possible. There were four different scenarios; no problems of any kind, failure of the aircraft's VOR indicator near a checkpoint, vacuum system loss, and ILS indicator failure. Each person faced only one such condition on any given flight.

The only difference between the simulator exercise and a normal flight was that the pilots were asked to verbalize what they were thinking and doing as they went about their business. This, along with the simulator data, was recorded and analyzed.

The researchers wound up identifying 16 different mental models used by the subjects, characterized by the sequence of actions and the relative amount of attention given to major flight functions.

Of the 16 models, three dominated. They were dubbed Model 1, where emphasis was on aircraft heading, followed by attention to the location/destination and flying the airplane; Model 6, a simple sequence of attention to heading, attitude, and flight; and Model 16, where prime attention went to basic flying tasks, followed by heading and location.

There was a difference in the mental models used by experienced and inexperienced pilots. The greybeards used primarily Model 6 (heading, attitude, flight), while the novices went equally heavily for Model 16 (primary attention to flight, followed by heading and location).

When measured against their performance, the model used by the more experienced pilots proved to be the one that produced the best performance. The outcome was less clear in the case of the novice pilots, where the preferred model produced decidedly mixed results compared to other approaches.

This suggests that beginners may still be finding their way in terms of determining the cognitive sequence that will yield the best outcome for the task at hand. This, of course, is entirely learning by doing, but such research offers a glimmer of hope that flight training might one day take into account the coping methods used by experienced pilots. Hopefully, these methods can be taught along with needle, ball, and airspeed.

Trouble Guaranteed

An accident investigator soon learns that a mishap is never the result of a single catastrophic event; there is always a string of things done or undone that causes the problem. When a pilot is able to recognize a potential causative factor and remove it from the sequence, the chain is broken and the day probably saved.

It's an entirely different story when a pilot is well aware of a problem with his airplane, and chooses to ignore it and press on. This is an excellent example of how judgment—or the lack thereof—ends up being the key causative factor in a mishap. In the incident that follows, a pilot whose Piper Comanche ran out of fuel during an ILS approach to the Charlotte International Airport at Charlotte, North Carolina, was fully aware of his potential fuel problems but chose to press on anyway.

The weather at Charlotte was generally 800 feet broken, visibility two miles in rain, so the pilot filed an IFR flight plan from his base in Spring Hill, Florida, to Charlotte. He departed Spring Hill at about 8:30 a.m. and proceeded north. The flight was uneventful until the pilot was on the final approach segment of the ILS runway 36R approach into Charlotte, when the engine quit at 11:45 a.m. It had run a total of three and a half hours since being refueled.

According to the pilot, he switched tanks, but was unable to get the engine started. The Comanche broke out of the clouds at between 500 and 600 feet AGL. By this time, the pilot had little choice of landing sites, and was forced to put the airplane down in an orchard. During the landing all three landing gear legs were torn off and both wings were damaged.

Investigators said almost no fuel was found at the accident site. The tanks were dry, as were the fuel lines and pumps. There was only half an ounce of fuel in the carburetor bowl.

The airplane had been topped off two days before the accident, then it was flown to the Spring Hill airport; this flight took only a quarter of an hour, leaving ample flying time for the Charlotte trip.

But that's not the complete story. According to the pilot, fuel would normally leak out of the thermos-type fuel caps in flight when the tanks were full. He also said that he had noticed fuel stains on the upper surface of the wings behind the filler necks before the flight to Charlotte, but nevertheless did not visually check the fuel supply before departure.

Looks like we need to add another item to the list of things that are useless to pilots; altitude above, runway behind, and fuel outside the tanks.

From Frying Pan to Fire

The aviation regulations require that a pilot load enough fuel for an IFR flight so that he can reach his destination with plenty of flying time left in the tanks. This rule applies to all flight operations, and is of course intended to provide for unexpected circumstances such as extended holding, below-minimums weather and so forth. It's a very good rule.

The accident reports occasionally describe a situation that on its face, appears to be one of complete disregard of the IFR fuel rule. In the report that follows, the pilot must have known that his fuel load was insufficient for the flight he undertook—this was an operation conducted under Part 135, which mandates a closer look at destination weather, and in this case, conditions clearly required an alternate airport. Here's what happened.

> The air taxi cargo flight was arriving at Jacksonville's Craig Municipal Airport from Panama City, Florida, at the end of an hour-and-fifteen-minute flight. The lone pilot made a VOR approach to Runway 13 in weather that featured a 100-foot ceiling, sky obscured, visibility one mile in fog. The pilot later told investigators that he actually touched down on the runway, but realized that he had landed too far along the runway to get the aircraft stopped, so he took off again.
>
> After becoming airborne, he began a right turn to keep the lighted area of the terrain under him, but at 300 feet AGL the engine stopped, and he was forced to make an emergency landing in trees near the airport.

When investigators arrived at the scene, one of the first things they did was to drain the airplane's fuel system, a job that didn't take long because the entire system contained only one pint of gasoline. Fuel was added, whereupon the engine started and ran flawlessly.

All the News That's Fit to Print

Aviation educators have always taught instrument pilots that if an approach procedure is flown precisely as charted, there is no way that obstructions will be encountered. Those who design the procedures and draw the charts take everything into account, but "everything" includes a lot of information, much more than could be printed legibly on those 5 by 8 sheets, and so the powers-that-be have no recourse but to include only the essential words, numbers and symbols.

Even with printing efficiency at work, the IFR approach chart is not always a marvel of clarity and simplicity. Compound the legibility problem with turbulence and perhaps poor cockpit lighting after dark, and you have a good reason for close scrutiny of the chart in calmer conditions before departure. A Piper Comanche pilot encountered this kind of situation early one February evening on an IFR flight from Portland to Newport, Oregon.

The 800-hour instrument-rated pilot was cleared for a VOR DME approach to Runway 16 at Newport, and in his own words, "I proceeded inbound to the 9 DME point and turned onto the DME arc. I experienced some difficulty holding the DME arc due to gusty crosswind conditions. As I approached the 344-degree radial, I turned inbound on a 164-degree heading and tracked the VOR, commencing my final approach.

"Due to the turbulent conditions, the bad weather, the night-like conditions and certain other factors, I apparently misread the approach plate and proceeded under the misunderstanding that I was entitled to descend to 980 feet when turning off the DME arc. I descended to 980 feet MSL on my altimeter and proceeded inbound.

"I soon experienced severe turbulence and felt a downdraft. Moments later, we saw a dark shape looming out of the area ahead which looked similar to the top of a dark thunderhead. Instants later, this took on the shape of a hill with pointed tree tops on it. Due to the virtual zero-zero visibility conditions, it was too late for any go-around. Instantly after the sighting, we began to feel the impact of tree tops striking the propeller, fuselage, wings, undercarriage, etc., and less than two seconds later the airplane came to rest near a small gravel road."

The aircraft crashed 11 miles from the airport, or two miles outside the nine-mile DME arc. According to an FAA investigator, the pilot told him that he had descended to the elevation posted for *four* miles from the airport instead of nine miles. At nine miles out, an altitude of 2,700 feet should have been maintained instead of 980 feet. The pilot said that he evidently misread a "4" as a "9" on the chart, and was at the wrong elevation.

But he had a complaint about Newport's altimeter setting given to him by Seattle Center. He maintained that the setting he received was incorrect, and compounded his altitude error by putting him 100 feet lower than a correct setting would have done. The FAA report maintained that it had no record of Seattle Center issuing an altimeter setting for Newport to the pilot prior to the approach.

Very few approach procedures are as simple as we'd like them to be. Intervening obstacles, radio coverage and other factors usually force the designers to mandate a couple of minimum altitudes on the way to the airport. Instrument pilots must learn to brief themselves on the various segments of an approach and, talking to themselves if necessary, be certain that they are aware of the proper altitude for the next segment before proceeding. In other words, always be one step ahead of the airplane. It's often too much to ask to memorize *all* that information, especially when there are more than a couple "step-downs" involved, but always take a few seconds to double-check the next minimum altitude before pressing on.

Workload Divided by Two

The pilot who strikes off into IMC all by himself has obviously shouldered the responsibility for all the chores that will be required. An autopilot is a big help, and another instrument-rated pilot in the right seat can ease the workload even more.

Here's the story of a rather complicated IFR operation in a Cessna 210 with a "crew" of two. The pilot in the left seat had 842 hours of flying time and an instrument rating, but that flight experience had been accumulated over many years, and he had logged only three hours of actual instrument time. Much more significantly, his most recent IFR time had been logged several years prior to the crash.

The pilot's wife, sitting in the right seat, was also a licensed pilot with an instrument rating. She was not considered pilot-in-command and the NTSB investigator did not retain her pilot records, so details of her flying time could not be obtained. However, she was believed to have logged somewhat less flying time than her husband, but had obtained her instrument rating relatively recently.

The couple set out on a trip from their home base in Colorado Springs, bound for Medford, Wisconsin; the pilot (husband) got a weather briefing and filed a VFR flight plan for the trip, although he apparently never activated it. The phone conversation with the FSS is of interest:

Pilot: I'm going VFR from Colorado Springs to Sioux City

Providing they're both competent and current, two pilots on the flight deck can substantially improve the safety of an IFR flight.

and then on to Wausau, Wisconsin. I'd like to have some weather in between here and there.

FSS: Ah, when are you planning on doing this dare-devil thing?

Pilot: Pardon?

FSS: When are you planning on doing this?

Pilot: Oh, just as soon as I can get briefed. The airplane's all ready to go, so it'll be 30 minutes.

The briefer went on to inform the pilot that VFR flight was not recommended because of a weak frontal system along the route of flight.

As they discussed the details, the briefer asked, "Are you IFR rated?" The pilot responded, "VFR—yes, I'm IFR rated and that's an IFR-equipped airplane." Discussing the weather further, he stated he could always go VFR as far as possible and "I'll just have to file from the air."

The briefer informed him that the restrictions in force

because of the ATC controller strike required at least an hour's lead time for IFR air-files. Faced with this news, the pilot said he would try to reach Sioux City VFR. He was about to hang up when the briefer suggested filing a VFR flight plan, which the pilot did, but no one heard him in fact activate it.

Three hours later, and still an hour away from his intermediate stop, the pilot contacted an FSS, stating he was on a VFR flight plan and going into Sioux City. He was given destination weather—2,100 broken, four miles in haze—and was advised that VFR was not recommended. His response was "Okay, sounds like we're going to have to make an approach."

In due course, he contacted approach controllers and requested a "local IFR...ILS approach." He was vectored onto the localizer, broke out well above ground, and conducted a visual approach into Sioux City, landing at 4:50 p.m. While taxiing in, the pilot inquired about filing an IFR flight plan to his final destination in Wisconsin.

At 6:13 p.m. the airplane was fueled and ready to go; the pilot was issued a clearance to VFR on top. He did so without incident, got on top above 7,300 feet, and canceled IFR.

He was not heard from again until 8:23 p.m., when he called the Wausau FSS, stating he was west of Wausau, requesting an IFR approach into Central Wisconsin Airport. The FSS again informed him of the restrictions on air filing of flight plans, but agreed to forward the pilot's request to Chicago Center, which would ultimately control the flight.

The FSS specialist picked up the land-line telephone linking him to Center (out of earshot of the pilot) and said, "Ah, I got a good one for ya," and he cooked up an odd flight plan. It proposed a "departure" at 8:35 p.m. from a point overhead the Wausau VOR to the destination airport almost directly below. This probably had no effect on the pilot, since when he called over the VOR it was 8:38 p.m. and he was promptly transferred to the approach controller and handled without unreasonable delay.

On the way to the Wausau VOR, the pilot was given the 7:55 p.m. weather observation for the field: Indefinite ceiling 500 feet, sky obscured, visibility one-half mile, light drizzle and fog, temperature 55, dewpoint 55. An airline pilot landing ahead of him also reported 500 and a half. The Centurion was now cleared for the ILS Runway 8 approach, for which the DH is 200 feet AGL and the visibility requirement one-half mile. The straight-in localizer approach also requires a half mile visibility, and the MDA is 368 feet AGL.

The pilot executed the initial approach and started down from 3,000 feet. Four minutes and four seconds later, a missed approach was declared; "We were having trouble with our localizer here," he told the controller, asking to return to the VOR to try again.

The controller checked the electronic monitors, which showed that the ILS equipment was operating satisfactorily. "Apparently, I've got something here on the panel," said the pilot.

So he tried another approach, and after two minutes the pilot said, "We got a couple of false (unintelligible) on the approach and that's what threw us off. Ah, now we finally got it nailed down. Why, no problem."

But five minutes later the pilot reported, "We had to make a missed approach, ah, not lined up and I, when I saw the runway I was too close." He headed out to the VOR for another try. In the meantime, the indefinite ceiling had dropped to 300 feet, visibility remained at half a mile. The pilot acknowledged the weather report and was not heard from again.

The wreckage was not found until the next day because of heavy fog. The crash site was off the far end of Runway 8, as if to suggest the plane was on yet another missed approach when it struck the ground out of control, steeply nose down. All aboard had been killed.

Investigators were disturbed at what they found. Not only were there discrepancies in the airplane's maintenance records—such as a failure to show compliance with a recent Airworthiness Directive—but it was clear the avionics of the vintage airplane had not been up to the task at hand.

They found records indicating that the plane's original glideslope receiver had been removed about a year and a half prior to the flight. It was replaced by a nav receiver, but one that did not have a glideslope. The glideslope antenna had been disconnected, and there was every reason to believe that the pilot knew he could not receive a full ILS signal. While that would not have prevented him from making a localizer approach, it would be incumbent on him to state the type of approach he was making.

Further, although the runway lights had been left on full intensity, it was necessary for the pilot to key his microphone to turn on the approach light system. Witnesses, including one who had monitored a scanner radio that night, said the approach lights were never activated during the time of the Centurion's approaches. Later in the evening, they worked when a corporate airplane flew the approach and landed.

This accident may seem to be the work of a pilot who believed he could overcome a total lack of recent IFR experience. But, although no one can know what went on in the cockpit that night, it would seem altogether a different matter if the pilot's wife, relatively competent and current on instruments, were overseeing her husband on the approaches.

Examining this possibility, however, one can see what a Herculean task she would have had. It would have demanded a constant monitoring of the instruments from a position in the cockpit that would encourage vertigo because of frequent head movements. While doing this, she would have had to try to soak up all the information on the approach plate, help with the radios, and more.

The approach chart for this procedure is not easily assimilated, and radio tuning could become a chore. For instance, the DANCI intersection—which serves as both IAF and FAF—is defined by Stevens Point VOR, but the missed approach intersection, HATLY, is defined by Wausau VOR.

If a pilot started from Wausau VOR, he would fly a course of 238 degrees until—assuming perfect navigation—he would pass over the outer marker, intercept the localizer and pick up the 318 radial of Stevens Point. Using the rudimentary localizer display on this pilot's panel, he would then have to track the localizer outbound by doing the opposite of what the needle says. These tasks could be very confusing for an instrument pilot with no recent experience.

Time to Share When You Go by Air

Time-sharing is generally thought of as a computer concept (or a condo-concept), yet there's probably no other ability that so clearly separates pilots of varying skill levels.

We know intuitively that those who can't seem to chew gum and walk at the same time will have trouble taking care of the multiple tasks that confront even a VFR pilot on an easy local flight in the daytime. Up the ante to night, cross-country flight, in marginal VFR conditions, and a considerable segment of the pilot population is challenged. When the name of the game is IFR, the ability to do more than one thing at once is not just a nice pilot option, it's a mandatory part of the flight kit.

All this suggests that some of the "right stuff" every pilot should have is the ability to time-share when confronted with multiple tasks. A researcher at NASA's Ames Research Center decided to take a closer look and see if pilots really do make better time-sharers. She selected 24 male college students, age 18-36, half were pilots, half were not. All the pilots except two were instrument rated, and all but one had either a commercial or instructor certificate. Total time ranged from 120 to 2,000 hours, with an average of 863 hours.

Half the people (divided evenly between pilots and ground-bounds) were placed in the "predictable difficulty" group, half in the "unpredictable difficulty" group. These designations had to do with the way in which certain changes in the assigned task were made.

The soon-to-be-besieged subjects were asked to conduct simulta-neous tasks that involved a tracking test with one hand while undertaking a "transformation" task, in the form of identifying the direction of a varying compass rose presented either visually or audibly, with responses being made either manually (via eight push-buttons) or by speech. This might be compared to trying to pat your head and rub your belly while discussing the economic implications of the decline in general aviation.

After much tracking and transforming, and measuring (and probably a fair amount of shrieking), the results were tallied. "In general," the researcher reports, "the differences observed between the students and the pilots have not been overwhelming."

It appears, concludes the scientist, that "the interactions be-tween the various resources are not easily altered by training and that experimental results obtained from college students may be applicable to the operators of real-world systems."

Based on this study, at least, it would appear to be a case of "you either have it or you don't." Perhaps there really *is* something to the concept that some pilots have "the right stuff."

A Practice Approach—And a Real Crash

There may be hundreds of examples of a pair of closely situated airports that have a special relationship in the minds of IFR pilots; Airport A is a well-lighted field with several ILS approaches and a tower open all night, and the nearby Airport B has a lower-quality lighting system, is not attended at night, and its best approach is a VOR-A or NDB. It often turns out that for reasons of convenience, Airport B is really where the pilot would rather be.

When ceiling and visibility permit, the answer for many pilots is to start down the ILS approach to Airport A, break out of the clouds and then go VFR to Airport B. There are many times when terrain and weather conditions allow this to be performed in reasonable safety. However, it can never be quite as safe as landing at Airport A and accepting the inconvenience.

The hazards of the IFR-to-VFR procedure were demonstrated on an April night in Michigan, when a Cessna 172M crashed just south of the Tri-City Airport at Saginaw. The pilot had successfully flown the ILS Runway 5 approach there, had the runway in sight and could easily have landed—but he chose to cancel IFR and attempt to go VFR to Harry W. Browne Airport, 11 miles to the southeast on the other side of the city.

In our opinion, this foolhardy act (which ended in a crash within a minute) was the crowning achievement of a pilot who seemed bent on demonstrating that he knew everything he needed to know about flying in weather—except how to heed advice.

The pilot had a total of 475 hours, a recent instrument rating, and an estimated 40 hours of IFR time. Much of his pilot experience was in the Skyhawk, which he owned jointly with several other pilots. The plane was based at Harry Browne Airport, an uncontrolled field on the east side of the Saginaw River.

At around 6 p.m. the evening of the accident, the pilot and a friend prepared to engage in some practice instrument flying. He called the local FSS and got a not-very-encouraging weather briefing; there was a front lying south and west of the area, with convective activity expected in the Saginaw area. "I recall my stressing the potential for such activity," the briefer later stated. "At first, the pilot gave me the impression that he wanted to remain in the

Improvement in ATC radar may soon make it possible for controllers to vector aircraft around small areas of weather that are often contributing factors in IFR accidents.

local area. However, in the course of the briefing, he requested information for a flight to Muskegon."

The briefer said he provided current and forecast weather for Saginaw and Muskegon, about 100 miles west. Saginaw had 500 broken, 1,200 overcast, visibility 5 miles in haze and was forecast to remain about the same, but with rainshowers and thunderstorms. Muskegon was 1,600 broken, 25,000 overcast, 3 miles in fog and haze, but forecast also to go down to 500 broken, with rain and thunderstorms.

"I also mentioned the likelihood of thunderstorm activity along his intended route, seeing that the Lansing, Jackson Country, Detroit and Pontiac areas were reporting thunderstorm activity at that time," the briefer recalled. "Also, there were two separate convective SIGMETs which skirted the area of his intended operation." All of this information was conveyed to the pilot as well as a statement to the effect that any single-engine, light-aircraft flight would be risky at best. The briefing terminated with no flight plan being filed.

About 40 minutes later (6:49 p.m.), the pilot took off from Harry Browne, and once in the air contacted Tri-City Approach, requesting a VOR Runway 32 approach, which was approved. As he began vectors for that approach, he informed the controller that it would be a practice approach, and that after completing it he would like a clearance to Muskegon at 4,000 feet. This was promptly approved by the controller as an air-filed flight plan.

The leg went without incident, and at Muskegon the pilot was cleared for a localizer approach. Once again, he informed the controller it would be a practice approach and requested a clearance back to Saginaw afterward.

The approach at Muskegon went without incident and the pilot was soon on his way home. It is worth mentioning that on the approach, the controller informed the pilot, "We have weather (i.e., a thunderstorm cell on radar) 15 miles northwest of Muskegon, moving eastbound at about 25 knots." The pilot responded, "Okay, ah, we'll be leaving as soon as we make the approach." En route, the pilot asked the controller, "Are you showing anything along Victor 216?" The controller said, "Not between you and VANNY (an intersection near the midpoint of the route). I show a pretty clean route."

The pilot routinely said, "Have a nice day" to every controller as he was leaving the frequency. Now we don't want to argue against good relations with controllers, but this kind of communication sometimes seems to be an effort by the pilot to impress the listener with his unflappable command of the flight—"Look at me. I'm not only flying on instruments at night, which almost makes me an airline pilot, but I have the time to be concerned about you earthlings down there."

Approaching VANNY, the pilot was handed off to Saginaw Approach, whereupon he told the controller, "What we want to do is probably shoot, ah, the ILS (at Tri-City) and then maybe, ah, we'll either proceed VFR over to Browne or we'll shoot the VOR-A approach into Browne." He had told the controller he had received ATIS information Mike ("Sky partially obscured, measured ceiling 200 broken, 600 overcast, visibility one and one-quarter mile in fog and haze").

The controller considered this request for several minutes and at about 8:44 p.m., there was this exchange:

Controller: Cessna 24R, I understand you want to land here and, ah, then pick up the approach?

Pilot: Ah, 24R, that's a negative. We'll, ah, do the low approach and, ah, when I get down to minimums I'll see what it looks like. Ah, maybe I can go VFR over to Browne, or else I'll shoot the approach into Browne.

Controller: Cessna 24R, roger. Unless the weather changes drastically by the time you get to Saginaw, I doubt you'll be able to go VFR. Ah, you said you had Mike, is that correct?

Pilot: 24R, that's affirm.

This particular controller did not pursue the topic further. At 8:53, however, he did transmit, "Cessna 24R, numerous scattered weather areas, 11 to 1 o'clock, 12 miles, intensities unknown. Largest one is a mile in diameter, moving slowly northeastward." The pilot replied, "Ah, 24R, thank you."

Vectors for the initial approach began. At 8:56, the controller remarked, "Cessna 24R, we just had a Navajo go through the scattered weather areas at 6,000. He said he was between layers with light rain." The pilot replied, "And 24R, we got light rain and that's about all."

This controller gave a final vector and cleared the pilot for the ILS Runway 5 approach at Saginaw once he was established on the approach course. When handed off to Tri-City tower, controllers assumed that the pilot would execute a practice approach there, then a VOR approach to Browne. The rest of the transcript tells what happened:

Controller: 5224R, Saginaw tower. Cleared low approach Runway 5. After your low approach, turn right, intercept the Saginaw 125 radial, fly it outbound for the Browne approach, climb and maintain 2,200.

Pilot: Okay, ah, 24R, we'll, ah, when we're coming down the ILS we'll see what the ceiling is and, ah, maybe we'll (unintelligible) want to go over VFR.

Controller: Well, I, ah will advise you now that they're— most of the approaches in here in the last hour have been, ah, marginal approaches down to minimums before they see anything.

Pilot: 24R, thank you.

Controller: 24R, you may, ah, find you should have to amend your plans and, ah, make a full-stop landing here at Saginaw due to the, ah, height of the ceiling.

Pilot: Yeah, ah, 24R. That's what I'm gonna check out on the way down.

Controller: Roger.

Pilot: Saginaw, 24R's gonna cancel IFR. We're gonna go VFR over to Browne.

Controller: Well, sir, the Saginaw control zone is IFR with a 200-foot broken ceiling.

Controller: 24R, we are IFR. Say intentions.

Controller: 24R, are you still on frequency?

There would be no answer—investigators believe the pilot crashed within a minute after his last transmission. The wreckage was found about 1,100 feet south of the middle marker, or about three-quarters of a mile south of the runway threshold. The Skyhawk had touched down, wheels first (i.e., probably wings-level but descending), then tumbled. Firemen arrived promptly, but the occupants had probably died of impact injuries.

The wreckage swath indicated a heading of about 115 degrees, or roughly the direction a pilot would take when breaking off the Tri-City approach and heading for Browne via the Saginaw VOR 125 radial. The No. 1 nav was still on the Tri-City ILS frequency, the No. 2 nav was on the Saginaw VOR frequency, and the OBS was on 120 degrees. These findings are all consistent with the pilot flying down the ILS to approximately the middle marker, breaking out and seeing the runway, then veering off toward Browne. It could also be imagined that he may have been setting the new OBS heading at or just before impact.

Since the altitude at the middle marker would be only about 200-300 feet, it can be imagined that in turning away from the lighted runway toward a dark sector to the southeast, the pilot may have had too little ground reference to detect a loss of altitude in the turn, and had only enough time to roll out of the turn by reference to the artificial horizon before striking the ground.

To underline the foolhardiness of the act, it is worth noting that even if the pilot had been successful in coming around to the new heading and starting toward Browne at a presumed altitude of 200-300 feet AGL, the numerous antennas and other obstructions along the route would have presented an extreme hazard.

What could drive a pilot to make such a gross mistake? As far as we could learn, his only disadvantage in landing at Tri-City was the need to arrange ground transportation to Browne where his car was parked, and at a later date shuttle the Skyhawk back to base.

But his attitude may have been the most important factor. The pilot had been given broad hints at least three times—the weather briefer as much as suggested that he not go; the approach controller hinted that attempting to go VFR to Browne would be unsafe; the tower controller not only suggested stopping at Tri-City, but very broadly hinted that attempting the VFR scud-run to Browne would be not only unsafe, but illegal.

But by getting to Muskegon, and braving the "weather at 11 to 1 o'clock," the pilot had demonstrated that the briefer was wrong. And by breaking out slightly above minimums on the ILS approach to Saginaw, he showed that the approach controller was wrong. He now had only to sneak over to Browne Airport to prove that the tower controller was wrong as well. Unfortunately for this pilot, the controller was right.

Index